T5-BCI-281

Lecture Notes in Biomathematics

Managing Editor: S. Levin

74

Andrej Yu. Yakovlev
Aleksandr V. Zorin

Computer Simulation in Cell Radiobiology

Lecture Notes in Biomathematics

Vol. 1: P. Waltman, Deterministic Threshold Models in the Theory of Epidemics. V, 101 pages. 1974.

Vol. 2: Mathematical Problems in Biology, Victoria Conference 1973. Edited by P. van den Driessche. VI, 280 pages. 1974.

Vol. 3: D. Ludwig, Stochastic Population Theories. VI, 108 pages. 1974.

Vol. 4: Physics and Mathematics of the Nervous System. Edited by M. Conrad, W. Güttinger, and M. Dal Cin. XI, 584 pages. 1974.

Vol. 5: Mathematical Analysis of Decision Problems in Ecology. Proceedings 1973. Edited by A. Charnes and W.R. Lynn. VIII, 421 pages. 1975.

Vol. 6: H.T. Banks, Modeling and Control in the Biomedical Sciences. V. 114 pages. 1975.

Vol. 7: M.C. Mackey, Ion Transport through Biological Membranes, An Integrated Theoretical Approach. IX, 240 pages. 1975.

Vol. 8: C. DeLisi, Antigen Antibody Interactions. IV, 142 pages. 1976.

Vol. 9: N. Dubin, A Stochastic Model for Immunological Feedback in Carcinogenesis: Analysis and Approximations. XIII, 163 pages. 1976.

Vol. 10: J.J. Tyson, The Belousov-Zhabotinskii Reaktion. IX, 128 pages. 1976.

Vol. 11: Mathematical Models in Medicine. Workshop 1976. Edited by J. Berger, W. Bühler, R. Repges, and P. Tautu. XII, 281 pages. 1976.

Vol. 12: A.V. Holden, Models of the Stochastic Activity of Neurones. VII, 368 pages. 1976.

Vol. 13: Mathematical Models in Biological Discovery. Edited by D.L. Solomon and C. Walter. VI, 240 pages. 1977.

Vol. 14: L.M. Ricciardi, Diffusion Processes and Related Topics in Biology. VI, 200 pages. 1977.

Vol. 15: Th. Nagylaki, Selection in One- and Two-Locus Systems. VIII, 208 pages. 1977.

Vol. 16: G. Sampath, S.K. Srinivasan, Stochastic Models for Spike Trains of Single Neurons. VIII, 188 pages. 1977.

Vol. 17: T. Maruyama, Stochastic Problems in Population Genetics. VIII, 245 pages. 1977.

Vol. 18: Mathematics and the Life Sciences. Proceedings 1975. Edited by D.E. Matthews. VII, 385 pages. 1977.

Vol. 19: Measuring Selection in Natural Populations. Edited by F.B. Christiansen and T.M. Fenchel. XXXI, 564 pages. 1977.

Vol. 20: J.M. Cushing, Integrodifferential Equations and Delay Models in Population Dynamics. VI, 196 pages. 1977.

Vol. 21: Theoretical Approaches to Complex Systems. Proceedings 1977. Edited by R. Heim and G. Palm. VI, 244 pages. 1978.

Vol. 22: F.M. Scudo and J.R. Ziegler, The Golden Age of Theoretical Ecology: 1923–1940. XII, 490 pages. 1978.

Vol. 23: Geometrical Probability and Biological Structures: Buffon's 200th Anniversary. Proceedings 1977. Edited by R.E. Miles and J. Serra. XII, 338 pages. 1978.

Vol. 24: F.L. Bookstein, The Measurement of Biological Shape and Shape Change. VIII, 191 pages. 1978.

Vol. 25: P. Yodzis, Competition for Space and the Structure of Ecological Communities. VI, 191 pages. 1978.

Vol. 26: M.B Katz, Questions of Uniqueness and Resolution in Reconstruction from Projections. IX, 175 pages. 1978.

Vol. 27: N. MacDonald, Time Lags in Biological Models. VII, 112 pages. 1978.

Vol. 28: P.C. Fife, Mathematical Aspects of Reacting and Diffusing Systems. IV, 185 pages. 1979.

Vol. 29: Kinetic Logic – A Boolean Approach to the Analysis of Complex Regulatory Systems. Proceedings, 1977. Edited by R. Thomas. XIII, 507 pages. 1979.

Vol. 30: M. Eisen, Mathematical Models in Cell Biology and Cancer Chemotherapy. IX, 431 pages. 1979.

Vol. 31: E. Akin, The Geometry of Population Genetics. IV, 205 pages. 1979.

Vol. 32: Systems Theory in Immunology. Proceedings, 1978. Edited by G. Bruni et al. XI, 273 pages. 1979.

Vol. 33: Mathematical Modelling in Biology and Ecology. Proceedings, 1979. Edited by W.M. Getz. VIII, 355 pages. 1980.

Vol. 34: R. Collins, T.J. van der Werff, Mathematical Models of the Dynamics of the Human Eye. VII, 99 pages. 1980.

Vol. 35: U. an der Heiden, Analysis of Neural Networks. X, 159 pages. 1980.

Vol. 36: A. Wörz-Busekros, Algebras in Genetics. VI, 237 pages. 1980.

Vol. 37: T. Ohta, Evolution and Variation of Multigene Families. VIII, 131 pages. 1980.

Vol. 38: Biological Growth and Spread: Mathematical Theories and Applications. Proceedings, 1979. Edited by W. Jäger, H. Rost and P. Tautu. XI, 511 pages. 1980.

Vol. 39: Vito Volterra Symposium on Mathematical Models in Biology. Proceedings, 1979. Edited by C. Barigozzi. VI, 417 pages. 1980.

Vol. 40: Renewable Resource Management. Proceedings, 1980. Edited by T. Vincent and J. Skowronski. XII, 236 pages. 1981.

Vol. 41: Modèles Mathématiques en Biologie. Proceedings, 1978. Edited by C. Chevalet and A. Micali. XIV, 219 pages. 1981.

Vol. 42: G.W. Swan, Optimization of Human Cancer Radiotherapy. VIII, 282 pages. 1981.

Vol. 43: Mathematical Modeling on the Hearing Process. Proceedings, 1980. Edited by M.H. Holmes and L.A. Rubenfeld. V, 104 pages. 1981.

Vol. 44: Recognition of Pattern and Form. Proceedings, 1979. Edited by D.G. Albrecht. III, 226 pages. 1982.

Vol. 45: Competition and Cooperation in Neutral Nets. Proceedings, 1982. Edited by S. Amari and M.A. Arbib. XIV, 441 pages. 1982.

Vol. 46: E. Walter, Identifiability of State Space Models with applications to transformation systems. VIII, 202 pages. 1982.

Vol. 47: E. Frehland, Stochastic Transport Processes in Discrete Biological Systems. VIII, 169 pages. 1982.

Vol. 48: Tracer Kinetics and Physiologic Modeling. Proceedings, 1983. Edited by R.M. Lambrecht and A. Rescigno. VIII, 509 pages. 1983.

Vol. 49: Rhythms in Biology and Other Fields of Application. Proceedings, 1981. Edited by M. Cosnard, J. Demongeot and A. Le Breton. VII, 400 pages. 1983.

Vol. 50: D.H. Anderson, Compartmental Modeling and Tracer Kinetics. VII, 302 pages. 1983.

Vol. 51: Oscillations in Mathematical Biology. Proceedings, 1982. Edited by J.P.E. Hodgson. VI, 196 pages. 1983.

Vol. 52: Population Biology. Proceedings, 1982. Edited by H.I. Freedman and C. Strobeck. XVII, 440 pages. 1983.

Vol. 53: Evolutionary Dynamics of Genetic Diversity. Proceedings, 1983. Edited by G.S. Mani. VII, 312 pages. 1984.

Vol. 54: Mathematical Ecology. Proceedings, 1982. Edited by S.A. Levin and T.G. Hallam. XII, 513 pages. 1984.

Vol. 55: Modelling of Patterns in Space and Time. Proceedings, 1983. Edited by W. Jäger and J.D. Murray. VIII, 405 pages. 1984.

Vol. 56: H.W. Hethcote, J.A. Yorke, Gonorrhea Transmission Dynamics and Control. IX, 105 pages. 1984.

Vol. 57: Mathematics in Biology and Medicine. Proceedings, 1983. Edited by V. Capasso, E. Grosso and S.L. Paveri-Fontana. XVIII, 524 pages. 1985.

ctd. on inside back cover

Lecture Notes in Biomathematics

Managing Editor: S. Levin

74

Andrej Yu. Yakovlev
Aleksandr V. Zorin

Computer Simulation in Cell Radiobiology

Springer-Verlag
Berlin Heidelberg New York London Paris Tokyo

Authors

Andrej Yu. Yakovlev
Leningrad Polytechnical Institute
Politechnicheskaya ul., 29, Leningrad 195 251, USSR

Aleksandr V. Zorin
Central Research Institute of Roentgenology & Radiology
Leningradskaya, 70/4, Pesochnij-2
Leningrad, 188 646, USSR

Translated from the Russian manuscript
by Boris I. Grudinko

Mathematics Subject Classification (1980): 60J85; 62P10, 68J10

ISBN 3-540-19457-6 Springer-Verlag Berlin Heidelberg New York
ISBN 0-387-19457-6 Springer-Verlag New York Berlin Heidelberg

Printed in Germany

Printing and binding: Druckhaus Beltz, Hemsbach/Bergstr.
2146/3140-543210

To our mothers, Maria and Lucia

TABLE OF CONTENTS

INTRODUCTION

The search for ways to overcome tumour **radioresistance** is a major problem of experimental and clinical radiation oncology. The difficulties involved in the attempts to solve this problem are a matter of common knowledge. In many a laboratory extensive studies are underway of factors determining tumour tissue response to irradiation and of methods for exerting directional effect upon those factors. Such studies have revealed **that**, at least at the cellular level, a considerable number of factors manifest themselves which are responsible for radiation effect [1] . Among those are: spatial heterogeneity of tumour cell population producing radioresistant cell reserves (hypoxic cells of solid tumours); differing radiosensitivities of cell life cycle phases; intrinsic dynamics of the processes of radiation damage and postradiation cell recovery; induction of proliferative processes in response to the death of some cells within the population; the stochastic nature of cell kinetics and complicated interaction between individual cell subpopulations corresponding to different tumour loci. Questions arise as to whether the researchers are now in possession of adequate means for interpreting experimental findings and clinical evidence and whether there are procedures for performing complex analysis and predicting specific tumour responses to various irradiation regimens and to combined antitumoral effects, taking into account the complexities of the phenomena under study. It should be admitted that the development of such means for interpreting the empirical data, for testing hypotheses and designing new experiments lags far behind the needs of present-day radiobiology (and radiotherapy) of tumours, being a specific problem that still awaits solution. This, presumably, accounts for a certain decline of interest in studying regimens of fractionated tumour irradiation, since it is unlikely that a purely empirical search for the optimal regime might be successful. Scepticism as regards possibilities of predicting individual tumour response to irradiation and chemotherapeutic drug effects by cell kinetic parameters [2] is likewise due to the lack of theoretical substantiation for the prognostic value of various kinetic characteristics. The same problems are also inherent in the cell population kinetics in normal tissues.

A more constructive approach to devising efficient means for describing and interpreting experimental data is the elaboration of a fairly realistic model of the processes under investigation. Historically, cell kinetics has proved a fertile field for the application of mathematical methods. Mathematical (analytical) models of cell kinetics and its disturbances resulting from irradiation and chemotherapeutic drug effects are discussed in numerous publications. Nevertheless, mathematical models of antitumoral effects have not become a reliable tool of research as, due to purely technical reasons, the difficulties associated with the "curse of dimensionality" stripped the more complicated variants of such models of their analytical advantages.

The past decade has seen the development of a different approach - the so-called simulation modelling of the dynamics of tumour cell populations or, in other words, direct computer-based simulation of processes occurring within a cell system before and after its exposure to different damaging agents. That approach does not require the use of sophisticated analytical methods, it makes it possible to describe more realistically the phenomenon under study and to develop the model more expeditiously, introducing necessary modifications in the functional scheme of the system under modelling. The said advantages of computer simulation over purely mathematical approaches to the problem outweigh, in our opinion, its shortcomings stemming from the need to employ the sampling method in analyzing the properties of the model. To date, not only a series of tumour cell simulation models have been proposed on the basis of utilizing programming languages commonly used in simulation practice, such as GPSS and GASP, but specialized languages have been developed, intended directly for simulating various processes underlying the organization of the cell cycle.

The purpose of this book is to demonstrate the possible ways of using simulation for exploring cell kinetics. The monograph dealing mainly with the authors' own experience in the field, the emphasis is laid on the effects of cell radiobiology. However, Chapter I also contains a brief review of the developments in devising specialized software for constructing simulation models of complex systems, as well as a survey of results of simulating the dynamics of cell populations of different tissues in vivo and in vitro. Also discussed at length are the feasible means of obtaining from simulation experiments information essential for resolving various problems of cell biology. The subsequent chapters serve to illustrate concrete applications of the model.

Chapters II and III describe the structure of a model for the in vitro kinetics of normal and irradiated cell populations. In addition the results of employing the model for the analysis of radiobiologic effects at cellular level are discussed. In Chapter IV certain approach to the simulation of controlled cell systems is considered, the epithelium of the small intestine of the mammals taken as a case in point. There again particular attention is given to radiobiologic applications, though some general properties of feedback-controlled renewing cell populations are investigated as well. Chapter V is of a more special character. Two examples are considered which show how simulation modelling may aid in solving some puzzling problems of the mathematical theory of cell systems.

The approaches to modelling cell kinetics described in the book are neither universal nor the only possible ones. At present only a few first steps have been made along the way of extensive utilization of computers in studying the complex laws governing the reproduction, differentiation and death of cells. Evaluation of the results achieved along that route is left to the reader. We hope the book will direct the attention of biologists and biomathematicians to simulation modelling as a tool for gaining insight into biological processes and will stimulate the interest among computer science specialists in developing new, more efficient means for the simulation of such processes.

In our work on the book considerable help was rendered to us by our associates and friends. First and foremost we wish to thank Dr. Yu.V.Gusev whose findings underlie the discussion in Chapter IV and, partly, in Chapter V of the monograph. We are also sincerely grateful to Drs. V.A.Guschin and S.A.Danielyan, as well as to our Bulgarian colleagues Drs. N.M.Yanev, S.T.Rachev and M.S.Tanushev for their constructive advice both in conducting the investigations and in preparing the manuscript. Preparation of the English edition would have presented to us insurmountable difficulties but for the expert assistance by Dr. B.I.Grudinko. The technical assistance of G.M.Pukhova is acknowledged. The authors are indebted to the reviewer of the manuscript for the very helpful comments and recommendations which they have gratefully taken advantage of in finalizing the manuscript.

REFERENCES

1. Barendsen, G.W. Variations in radiation responses among experimental tumors, In: Radiation Biology in Cancer Research, Raven Press, New York, 333-343, 1980.
2. Tannock, I.F. Cell kinetics and chemotherapy: a critical review, Cancer Treat. Rep., 62, 1117-1133, 1978.

I. SIMULATION IN CELL POPULATIONS KINETICS

1.1. General Features of Simulation Technique

Methods for mathematical modelling of the processes of cell proliferation and differentiation in normal and tumour tissues and their practical utilization are the subject of a considerable body of works. The theory of cell population dynamics and its applications has now become a full-fledged division of mathematical biology. The fundamentals of the mathematical theory of cell systems have been more comprehensively covered in monographs [13, 25, 70]. According to some authors [12, 13, 20, 67] the following are among the currently central problems of mathematical modelling in that field:

(1) development of effective methods for processing experimental data on the dynamic behaviour of normal cell populations. These activities aim at identifying parameters essential for understanding and interpreting specific properties of a given cell system;

(2) quantitative description of changes in the dynamics of cell populations due to the effect of damaging agents;

(3) optimization of the regimens of therapeutic action on cell systems.

The latter problem arises most often in connection with attempts to improve the existing methods of radio- and chemotherapy of tumours. Recently a tendency has manifested itself for extensive use of simulation methods in solving the above problems. However, mention should be made of the differences between researchers in understanding the essence of simulation technique.

Thus, according to Shannon [58] simulation is a process of constructing models of a real system and of staging experiments on that model for the purpose of either understanding the behaviour of the system or evaluating (within the framework of limitations imposed by certain criterion or a set of criteria) different strategies maintaining the operation of the system.

In a review by Aroesty et al. [3] simulation of biological objects is taken to mean the use of various computer models, i.e. numerical computer-based studies of analytical and functional-logical models both of deterministic and probabilistic types.

Yakovlev [78] points out that simulation understood as a controlled experiment with a computer-realized model of a system is not infrequently regarded as a method for operation research. The report of the 3rd USSR Conference on Operation Research [40] states that in the Soviet literature on the exploration of complex systems it is assumed that, as a rule, a simulation model is not intended for stating and solving problems of mathematical programming. However, simulation models oriented to optimization problems come into being in increasing frequency. An independent line of investigation is represented by simulation experiments with controls specified by experts or according to plans whose synthesis is accomplished by means of the experimental design theory. Simulation experiments are taken to mean those in which the process under investigation is reproduced by means of a model with different alternative controls and with the subsequent analysis of the results. The authors of the report arrive at the conclusion that interpretations of the term "simulation" are basically similar in the works of different authors, though generally speaking the term "simulation model" is not quite appropriate since the attribute "simulation" denotes, in effect, the way of using the model rather then its type. The definition of "simulation" given by Kindler [29] and stemming from that of Dahl [8] is, in our opinion, more adequate and it corresponds fully to the concepts realized in the models dealt with the present work. That definition regards simulation as a method of research based on replacing the dynamic system under discussion by its imitator with which experiments are conducted, aimed at obtaining information on the system concerned. Reasoning from that definition, it should be recognized that the overwhelming majority of computer models of biological systems and processes are not of the simulation type although they show some attributes of the latter. Analysis of the applications of simulation technique to the study of cell systems enables formulation of the basic principles for constructing simulation models in that field.

Construction of a simulation model for a cell population involves, as a rule, the following stages:

(1) development of functional diagram representing quantitative regularities and logical relationships characteristic of the cell system under investigation;

(2) realization (formalization) of the diagram. In the literature one comes across models which, for describing cell systems, use various analytical approaches, as well as different ways of imitating directly the object of research by means of digital and analog computers;

(3) elaboration of the procedure for employing the model which consists in the use of a technique for simulation model trials. Most of the authors conduct computer experimentation, i.e. computer-based simulation of the model's evolution with time and/or space. For ins-

tance, in employing stochastic models of cell populations, use is commonly made of statistical simulation methods.

1.2. Principal Trends in the Development of Methods for Stimulating Cell Proliferation Processes

Simulation as a method for studying the dynamics of cell populations appears to have been first employed in 1956 by Hoffman et al. [23]. The authors turned to the Monte-Carlo method for modelling the growth of a homogenous cell population of which each cell has a mitotic cycle duration generated in accordance with the specified distribution function. Further development of cell population models followed accumulation of knowledge concerning temporal organization of the life cycle of cells. On the basis of data obtained by radioautography some models were constructed which included a more detailed description of the processes occurring in the cell in the course of the mitotic cycle.

Most of the existing models of cell systems belong to the multicompartmental class. In effect, models of that class use the system approach which considers a complex biologic system as a set of interacting subsystems.

One of the frequently used schemes of detailing a homogenous cell population structure was proposed by Skipper [61], the block scheme including six states of a cell (Figure 1.1). States 1-4 in the figure correspond to phases of the mitotic cycle (MC), state 5 - to the

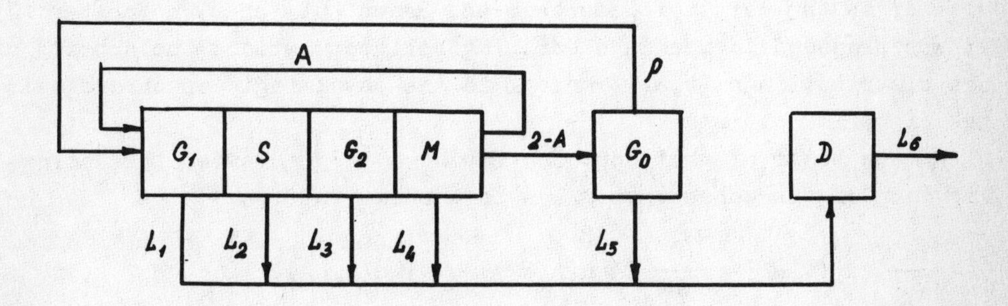

Figure 1.1. Functional block diagram for Skipper's model [61] . G_1, S, G_2 and M - the mitotic cycle (MC) phases; G_0 - the resting state relative to proliferation; D - a population of irreversibly differentiated cells; L_1 - L_6 - functions of cell death in the states in question; A - the quantity determining the fraction of cells which reenter

the MC, bypassing the G_o state; β - the fraction of nonproliferating cells entering the MC.

nonproliferating cell population, and state 6 - to that part of the cell cycle in which maturation and terminal differentiation culminating in cell death take place. L_i functions ($i = 1,\ldots,6$) characterizing the cell death rate were introduced for each state. The quantity $A \in [1, 2]$ determines the portion of cells reentering the MC; β is the portion of nonproliferating cells entering the MC. Distribution of cycle phase durations was assumed to be known in Skipper's work [61] ,but the duration of mitosis was regarded as deterministic.

That model was used by the authors in describing the development of a solid tumour and afterwards served as the basis for devising most of the models of homogeneous and heterogeneous cell populations [12, 38, 68].

1.3. Simulation Models of Cell Kinetics with a Deterministic Structure

Deterministic models of cell population kinetics play an important part both in eliciting general phenomenological regularities and in analyzing regulatory mechanisms functioning at cellular and molecular levels. Deterministic approach was used primarily in the wide class of the so-called age-dependent models. With that approach to the description of cell populations it is the differential form of the law of cell streams conservation that serves as the basic equation. Already in the works of Scherbaum and Rasch [52] and Von Foester [71] it was proposed to use in describing cell populations the density of age distribution $n(t,\alpha)$ where α is the physiologic or chronologic age of cells.

On the basis of that approach Rubinov [48] advanced the following form of the conservation law for density $n(t,\alpha)$

$$\frac{\partial}{\partial t} n(t,\alpha) + \frac{\partial}{\partial \alpha}(\nu n(t,\alpha)) = -\mu n(t,\alpha), \qquad (1)$$

where $\nu = d\alpha/dt$ is the cell maturation rate and μ is the rate of cell death, $\alpha \in [0,1]$.

In that case the boundary conditions reflect the contribution of splitting :

$$n(t,0)\nu(0) = \beta\, n(t,1)\nu(1),$$

where $\beta \in [0,2]$ is the birth-rate coefficient.

Equation (1) is also supplemented with the initial condition, i.e. $n(0,\alpha)$, initial age distribution, is prescribed. In effect, Rubinov made use of a continuity equation of the following type

$$\frac{\partial}{\partial t} n(a,t) + \mathbf{div}(\nu n) = -\mu(a,t)n(a,t) ,$$

which was also proposed by Oldfield [43] for the description of the progress of cells through the life cycle. In the opinion of some authors [3, 51] the main advantage of that approach is in that the growth of cells is represented by a continuous process. This enables description of changes in maturation rate in different parts of the mitotic cycle as well as modelling of different drug effects on cells. More detailed description of cell systems was made possible by introducing the function of joint cycle-duration and maturity distribution density $n_T = n(\alpha,T,t)$ [3, 7, 33, 57].

In that case equation (1) takes the form

$$\frac{\partial}{\partial t} n_T + \frac{1}{T} \frac{\partial}{\partial \alpha} n_T = -\mu(\alpha, T, t)n_T .$$

Investigation of the properties of complicated variants of age--dependent models appears impossible without the use of computers. For that reason some authors turned to computer simulation of such models. Thus, a discrete variant of an age-dependent tumour cell population model was proposed [3]. The functional block diagram of the model includes a set of delay lines for the number of cell clusters with different values of the mitotic cycle duration. Each line, in turn, is divided into a certain number of blocks according to the discrete values of cell maturity ($\alpha \in [0,1]$). The progress of cells through the life cycle is accomplished by generating time values for residence in each of the blocks (the Monte-Carlo method). That model was applied to the description of a population of solid tumour cells. Comparison with experimental data revealed fundamental conformity of the process of model system development to certain stages of tumour growth. Simulation of the results of chemotherapeutic treatment of the tumour showed the need for a more detailed model. In particular, it proved to be necessary to describe the dynamics of regulatory mechanisms which modify tumour response to a damaging effect.

The structural basis for simulating cell proliferation processes can often be provided by deterministic models which use the apparatus of conventional differential equations [45].

In the general form the process of changes in the size of a population is described by the set of equations :

$$\dot{x}(t) = Ax(t) , \quad x(0) = x_0 ,$$

where $x = \{x_i\}$ is the state vector and A is the population matrix.

By way of illustration let us consider a model of development and degeneration of the thymic gland [31] . The dynamics of individual thymocyte subpopulations is described within the framework of the model by the system of conventional differential equations derived by the use of the following considerations.

It was assumed that a certain proportion of bone marrow stem cells divide into two stem cells as expressed in the equation

$$\dot{L}_1 = 2 [1 - (L_1/S_m)] G_1L_1 ,$$

where L_1 is the number of stem cells, S_m is the maximum number of stem cells allowed, G_1 is the reciprocal of generation time for the stem cells. The rest of the stem cells undergo asymmetric mitosis resulting in one stem cell and one lymphoblast. The dynamics of the lymphoblast population is described by the equation

$$\dot{L}_2 = \frac{L_1}{S_m} G_1L_2 - G_2L_2 ,$$

where L_2 is the total number of lymphoblasts and L_2 is the reciprocal to their life span. The maximum number of stem cells S_m is a constant for the first fifty population doublings after which this value gradually decreases according to the equation

$$\dot{S}_m = - DG_1S_m ,$$

where D is the rate of decrease of S_m (degeneration parameter).

Once the lymphoblasts migrate to the thymus they begin a series of cell divisions associated with maturation processes and the following system of equations may be introduced

$$\dot{L}_{j+1} = 2G_jL_j - G_{j+1}L_{j+1} , \quad 2 < j < n-1 ,$$

where L_j is the number of cells in the jth generation, G_j is the reciprocal of the generation time of the jth generation. The last generation L_n (mature small lymphocytes) is described by the equation

$$\dot{L}_n = 2G_{n-1}L_{n-1} - 0.017L_n \ .$$

Similarly, for the kinetics of endothelial cells the following equation is used

$$\dot{E} = 2G_eE \ ,$$

where E is the total number of endothelial cells and G_e is the reciprocal of their generation time. The value G_e is function of the net change in the total population of the thymus as expressed by

$$1/G_e = [N_pT_k/(T_k - T_{k-1})] + 50 \ ,$$

where N_p is the nutrition parameter and T_k is the total number of thymocytes on the kth iteration of the model. The last expression ensures that the endothelial cells will divide no faster than a generation time of 50 hrs. Irradiation was simulated in the work [24] by simply reducing the number of cells in each generation by the prescribed proportion.

The system's trajectory was calculated numerically on a computer, and qualitative conformity of the model's behaviour to the real processes of development and degeneration of the thymus was established. However, the author's attempt to describe within the framework of the model the process of radiation injury of the thymus may be classified only as a first approximation since the process of radiation injury of the cell is represented as a one-stage event and no account is taken either of the process of postradiation cell regeneration or of the operation of mechanisms controlling repopulation of the thymic tissue.

Considerable progress has been made in the development of models of cell population dynamics with a discrete set of states. Dealing with that type of models is a series of studies [64, 65, 73] in which the process of changes in the numbers of cells differing in chronological age is approximated by a step function with the definite constant step Δt. The state of the population is described by the finite vector $C(t) = \{c_i(t)\}$, where $c_i(t)$ is the number of cells that entered the MC in the time interval $(i-1)\Delta t$, $i=1,\dots,k$. Variation of the vector with time is described by the matrix equation:

$$C(t+\Delta t) = TC(t)$$

where T is the matrix operator of the system.

The model is based on the use of the population projection or Leslie matrix which is well known in population biology. The simulation of cell population dynamics requires the use of finite, discrete distributions of transit times for cell cycle phases; continuous distributions may, however, be adequately approximated with this model. An expression (and its asymptotic behaviour at $t \to \infty$) was obtained for the fraction of labelled mitoses in the case of a homogenous cell population.

The method actually used in the above-mentioned studies [73] was that of simulation modelling since, despite the existence of an analytic description of the system concerned, the evolution (with time) of the cell ensemble was reproduced by means of a computer.

1.4. Stochastic Simulation Models in Cell Kinetics

Another line of development of the methematical theory of cell systems is concerned with the construction of stochastic multicompartmental models. Probabilistic description of cell kinetics is, as a rule, associated with well studied models of random processes: the Markov processes with continuous time and a discrete set of states [46, 47, 59] , as well as with continuous time and a continuous set of states [4, 17, 21, 28] , age-dependent branching processes [25, 39, 74, 75, 76] and renewal processes [53, 54, 77] . Stochastic models make it possible to study, in addition to average characteristics of the process, the part played by random fluctuations.

It should be noted that some authors believe that the results of stochastic and deterministic descriptions of compartmental systems are, identical. Indeed, it has been shown [14] that for a certain class of models there is a one-to-one correspondence between those methods of description, i.e. for one and the same closed multicompartmental system there is a one-to-one correspondence between the parameters of the deterministic model and its stochastic counterpart. In that case the expression which defines the behaviour of the average characteristics of a stochastic system coincides with the corresponding deterministic equation. However, Feller [16] has demonstrated that in the event of cell deaths in a growing cell population (an open compartmental system) the expected characteristics of a random process no longer coincide with the corresponding characteristics of the deterministic analogue. Even for the simple birth-death stochastic process there is a nonzero probability of population extinction within finite time. Clearly, that property cannot be established by studying the behaviour of the deterministic analogue of the birth and death process.

Further progress in the stochastic theory of cell systems is associated with the development of the model for the multiphase birth-death process [27]. It was precisely this trend that provided the basis for the construction of simulation models of stochastic systems and their practical application to cell kinetics.

That approach has made it possible to obtain [63] the following system of equations for the expected number of cells in the MC phases on the assumption of the independent transit of each cell through the life cycle :

$$\dot{N}_1(t) = 2\lambda_k N_k(t) - (\lambda_1 + \mu_1)N_1(t) ,$$

$$\dot{N}_i(t) = \lambda_{i-1}N_{i-1}(t) - (\lambda_i + \mu_i)N_i(t) , \quad i=2,3...k ,$$

where N_i is the expected number of cells in the ith phase of the MC; λ_i is the rate of transition from the ith phase to the next; μ_i is the death rate in the ith phase. Numerical solution of that set of equations was obtained with computer assistance. This enabled investigation of the effect of changes in the parameters λ and μ on the shape of the simulated fraction labelled mitoses curves.

Whenever there was a large number of subphases, a diffusion equation with a constant transfer coefficient was used as the structural basis for the simulation model [3] . Construction of a refined variant of the model in which probabilities of intersubphase transitions depend upon the phase position within the mitotic cycle leads to a diffusion equation with nonhomogeneous transfer and diffusion coefficients [3] . That approach was further advanced with the transition from solving the limited problem of identifying experimental kinetic curves to describing the dynamics of complex cell populations under various experimental conditions. For that purpose considerable sophistication of model representations was required with a view to allowing for the mutual influence of cells in the course of vital activity and for the heterogeneity of the cell composition of tissues. Owing to greater dimensional representation and to allowance for nonlinear effects analytical description and investigation of resulting stochastic models of cell systems have become quite laborious. It is for this reason that analytical description of heterogeneous populations has not been achieved in the framework of stochastic approach [47] .

Some authors, omitting the stage of analytical description, are devising models in which dynamic regularities of the behaviour of a real cell system are reproduced directly on a computer. Development of such models went on through refinement of compartmental schemes aimed at the maximum possible representation of actually observable properties of concrete cell populations. Most often the subject of research in such works was tumour cell populations [10, 38, 67] , cells in vitro [1, 60] and populations of stem cells of certain tissues [34, 42] . A typical example is the model developed by Valleron and Frindel [69] which falls into the category of simulation models and is an original computer program used by the authors for simulation of radioautographic experiments and instestigation of indices of cell kinetics. The model can be employed for tracing the fate of an individual cell and its progeny over a specified interval under fairly general assumptions but it cannot be applied to the study of a whole cellular ensemble and, correspondingly, to the evaluation of the role of interaction between individual cells and cell subpopulations.

The simulation model for the development of chloroleukemia proposed in [68] uses the functional scheme (Figure 1.2) which is a particular case of Skipper's scheme (Figure 1.1). The scheme includes four states of the cell (G_1, S, G_2, M) and specifies distribution of cells by the time of residence in the G_1, S and G_2 phases. The dura-

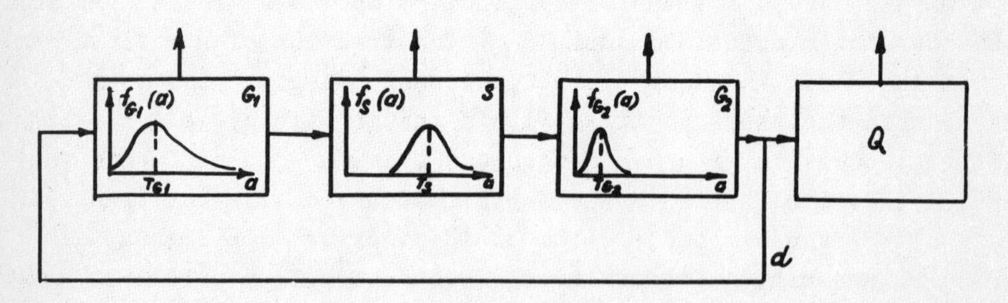

Figure 1.2. Functional block diagram for Toivonen and Rytömaa's model of chloroleukemia [68] . Q - a population of nonproliferating malignant cells; α - the fraction of cells reentering the mitotic cycle following division. Other designations as in Figure 1.1.

tion of mitosis is neglected. The time of residence in the Q-phase is unlimited (nonproliferating cell compartment). The introduced parameter α defines the portion of cells reentering the cycle after division. The transition probabilities $f_{i,i+1}$ for the time Δt were calculated by the following formula:

$$f_{i,i+1} = \int_{a}^{a+\Delta a} f_i(t)dt \;/\; \left[1 - \int_{0}^{a} f_i(t)dt\right] \quad ,$$

where Δa is the change in the cell age in the ith phase within the time Δt, and $f_i(t)$ are the prescribed densities of the distribution of the ith phase duration. The model was realized on a computer using the FORTRAN-IV language. As a result the following data were obtained at each step of simulation (Δt); the total number of cells in each phase, the growth fraction, the number of labelled cells in each phase, the fraction of labelled mitoses and the number of cell deaths within the given time interval. In simulation experiments the authors traced the development of a population of 350 to 10^4 cells. The model parameters were prescribed on the basis of available experimental data on chloroleukemia in man. Investigation of the effect of cytostatics on the behaviour of different cytokinetic indices was the purpose pursued in developing the model. It should be noted that the absence of the resting state for tumour cells with regulated entry into the MC restricts considerably the field of application of that approach, confining it to homogeneous exponentially growing cell populations. No real problems of investigating therapeutic effects on a tumour population were raised in that study. Applications of the model were limited to studying the effect of certain disturbances of cell kinetics on the form of the labelled mitoses curve, i.e. a problem earlier resolved by other means [36] . The scheme in Figure 1.2 appears to us too simple to justify, in this particular case, the application of simulation modelling as a method of research. It does not take account of many, now familiar regularities in the processes of mammalian cell proliferation and in the regulation of such processes.

The dynamics of a leukemic cell population was interpreted from a more general standpoint in terms of the queueing theory[38] . In that study the process of numerical changes in the system was formalized as an integer-valued discrete in time stochastic process. It determined the authors' choice of the technique of discrete modelling using the GPSS/360 specialized simulation language. In the program written in that language each cell is assigned a corresponding transaction (transactions are parameter arrays defining the principal characteristics of the cells being simulated).

For all MC phases (except the M phase) distributions of their durations were specified (Figure 1.3). The transit time in the M phase was assumed to be deterministic. According to the model after divi-

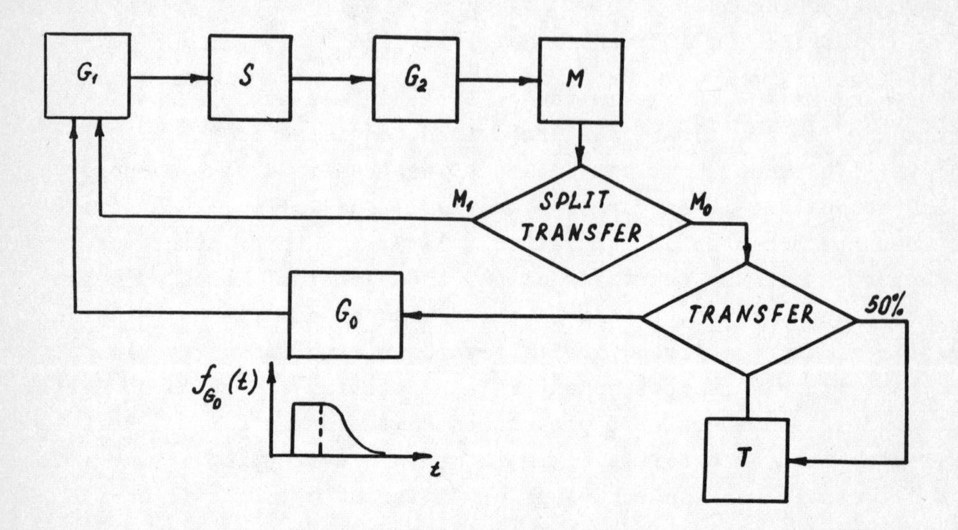

Figure 1.3. The model of a leukemic cell population by Mauer et al. [38] . G_0 - the resting state relative to proliferation; T - the cell death block.

sion 20% of cells reenter the MC (M_1 = 0.2) and 80% pass into the resting state G_0 (M_0 = 0.8). The residence time in the G_0 state was specified by a special type distribution density truncated on the left (Figure 1.3). In the block scheme (Figure 1.3) the maximum time of delay in G_0 (200 hours) was also indicated. The distribution of cells between the MC and the G_0 state was accomplished in the TRANSFER block. Cell division in the model was simulated by the splitting of a transaction (the SPLIT block in the scheme) resulting in two identical copies (symmetrical mitosis). Cell death was simulated by eliminating transactions from the system at the required place and specified rate.

Pre-assigned as initial conditions was the number of cells in each phase at the zero instant. The authors pointed out that once the initial conditions were specified the model was able to function in the "selfmaintenance" regime, reproducing on a computer the process of development of the cell system. The model was used for carrying out the following simulation experiments:

(1) Investigation into the effects of cell death rate on the growth

rate of a leukemic cell population. It was shown that the growth stopped at a death rate equal to 62.5% of the number of cells entering the G_0 state.

(2) Investigation into the effect of changes in the proportion of cells entering the G_0 or G_1 phases on the population growth rate.

(3) Simulation of the process of growth from a single cell to a clinically distinct population (10^{12} cells) Experiments have shown that a tumour requires some 5 years to reach critical dimensions.

(4) Investigation of specific effects of cytostatics in the case of vincristine. On that occasion the model structure had to be modified by including the death of cells before entering mitosis. The experiment illustrated the authors' contention that one of the advantages of their approach is flexibility of the model, possibility of making timely amendments, preserving at the same time the basic structure of the model. After the said modification the experiment yielded the labelling and mitotic indices under the effect of vincristine, their values being in agreement with clinical observations.

Among the shortcomings of the model [38] are :

(1) The duration of the S phase of the mitotic cycle is considered as a deterministic value equal to 20 hours. The need for such simplification is hardly accountable, seeing that the durations of the other phases of the mitotic cycle are assumed to be random variables;

(2) The fluxes of cells reentering after division the mitotic cycle or passing into the resting state are fixed. In effect, those fluxes are regulated quantities depending both on the external conditions and the total number of cells in (or density of) the population at a given moment.

(3) According to the model cell death occurs only in the G_0 phase and is modelled as an instantaneous process.

(4) The effect of cytotoxic agents is assessed by changes in the total number of cells in the population, whereas the most common method for estimating such effects is determination of the clonogenic capacity (survival rate) of damaged cells.

Despite the above-mentioned limitations the model of Mauer et al. [38] is, presumably, the only one described in the literature which enables simulation of simultaneous evolution of all cell ensemble elements.

In addition to refining the description of injury and regeneration of a cell system, the trend manifests itself in the literature towards description of cell population heterogeneity. The recent subjects of inquiry have been concrete cell systems in organized tissues, as well as heterogeneous tumours. However, a number of authors [47, 67] note that at present no detailed information is available which would enable construction of an absolutely correct model. Moreover, as regards complex differentiation processes there is no adequate basic functional scheme covering them. This has given rise to doubts as to the possibilities of the mathematical simulation of heterogeneous selfregulating populations. Thus, opinion has been expressed [47, 67] that regulation of cell proliferation remains an open area and that available data are a unquestionably insufficient for devising anything but speculative models.

It should be noted, however, that most of the investigators in that field are less sceptically-minded. Toivonen [67] for one believes that the lack of detailed information should not discourage a researcher. On the contrary, rigorous and realistic models should be developed predicting the behaviour of a population under conditions for which no experimental findings are at hand. Verification of predictions in real experiments would ensure development of the basic principles underlying the model.

In simulating renewing cell systems acceptance has been gained by Jansson's scheme [26] . In that compartmental model the process of differentiation is discrete, the principal maturation stages differing in the ability of cells to proliferate are distinguished, and the rates of transition between subpopulations are prescribed. Generalized for the case of probabilistic description of the processes of tissue renewal, this approach was used extensively in simulating differentiation pathways in hemopoieses [5, 22, 34] .

Based on different assumptions is the consideration of the process of cell differentiation in some studies [7, 41] using the concept of continuity of the process of hemopoietic stem cell maturation. The literature on simulation modelling of hemopoietic tissues is quite voluminous and deserves a detailed discussion in a separate review. Among the more important studies in that field, in our opinion, are [5, 30, 35, 37, 44, 49] .

Besides the hematopoietic system, another noteworthy subject of inquiry is epidermis. The simulation model developed by Honda [24, 50] is intended for studying features of the spatial organization of renewing tissue on the basis of analytical description of the form of epidermocytes and the prescribed algorithm of cell movement along the route of differentiation. The model demonstrated the stability of epidermal columns resulting from multiplication of basal layer cells.

Heterogeneity of tumour cell population was taken account of in simulating the development and regeneration of solid tumours [10, 11, 12] by considering a mixture of normal and transformed cells. The life cycle of tumour cells is represented in the model by a modification of Skipper's scheme (Section 1.2). In the studies by Düchting et al. a two-dimensional model [10] and afterward a three-dimensional one [11] were constructed, the latter enabling reproduction of the spatial structure of a developing tumour with consideration for heterogeneity of cell composition and the effect of the blood capillary system.

Thus, the method of direct cell system simulation is finding ever-widening application in studies on heterogeneous population both of normal and tumour tissues.

1.5. Simulation Modelling Software

To date over 500 simulation languages [29] have been developed for different problems and types of computers. They may be divided into three principal classes: languages for continuous, discrete and combined modelling. The latter type of languages is designed for simulating systems and processes which manifest simultaneously continuous and discrete features.

Languages of the first type (for continuous modelling) make up quite a large group and are mainly designed for simulation of systems described by means of differential equations. The use of such languages for simulating cell systems is expedient wherever there is a model in the form of a block-oriented or link-oriented scheme described by a set of dynamic equations. In that case the use of simulation languages merely simplifies the process of program realization of numerical calculations but, conceptually, it implies no innovation in modelling biological and, specifically, cellular systems. Such languages are typified by block-oriented DIGSIM [2, 56] and link-oriented DYANA [66] .

Discrete simulation languages are divided into several groups intended for solving specific types of problems. Thus, there are languages devised especially for designing and studying characteristics of computers and computer nets, although these problems may be resolved by means of other languages as well.

A large group of discrete simulation languages incorporate in their structure a transaction, a discrete element capable of being assigned a meaning varying with the nature of the problem. Many transaction languages are designed for simulating queueing systems. Transactions are interpreted to fit the system being simulated and may denote a shopper in the store, a bus in the transportation system, a living cell and a great variety of other objects.

The discrete simulation languages most generally employed are, apparently, SIMULA, SIMSCRIPT, GASP and GPSS. Two more specialized languages CELLSIM [9] and CELLDYN [15] designed for simulating cell systems should also be mentioned here. They will be discussed in some detail in what follows.

Selection of the type of language for model realization is determined by several factors, such as the researcher's skills, the purposes of simulation and, finally, the adequacy of the language for describing the specific processes and systems. The use of discrete simulation languages appears to be quite advantageous in studying a cell system (a cell culture or organized tissue). With this approach the life cycle of each individual cell is simulated with consideration for its peculiar features and the random character of reactions to different effects.

Whenever no individualization of cell properties is required, use may be made of continuous simulation languages. The functional scheme of the system under study is represented, in that case, as a set of compartments characterized by the number of cells within them, the rules governing transitions between them and the values of transition rates.

In some cases, when the system being simulated consists of cell populations differing in size by several orders or else the minimum number of cells required for modelling exceeds the capacity of the discrete language, transition to continuous or combined languages is expedient.

Combined simulation languages are present-day developing means for realistically describing complex systems in which it is desirable or necessary to take into account several subsystems some of which are of continuous while others are of discrete character. For a special study on simulation languages, introduction of more rigorous and comprehensive definitions is essential, specifically for combined simulation languages. Here the above characterization seems sufficient.

The languages in question may find extensive application in simulating the effects of damaging factors on regulated complex biological systems. An example will be given in the present book of a system where the possibilities of GPSS are still adequate (Chapter IV).

However, even for simulating postradiation dynamics of hemopoiesis, individualization of cell precursors' behaviour in the bone marrow is essential and, therefore, the use of discrete modelling concepts is quite appropriate. On the other hand, in describing hemopoiesis, account should be taken of a very great number of peripheric blood cells of different types. For that part of the hemopoietic system continuous simulation languages are amenable. Inasmuch as the "continuous" and "discrete" compartments are in interaction, i.e. there are intricate regulatory relations between them, combined simulation languages would be most adequate for modelling systems typified by hemopoiesis.

In using discrete languages for certain models of complex cell systems, the principal limitations are the number of elements (cells) whose life cycles are simulated simultaneously and the time of simulation which increases proportionally to the number of cells. These features are quite natural, being inherent in all discrete languages and, apparently, they should not be regarded as shortcomings.

The problems of simulating cell systems may be solved by means of plentiful existing software. Nevertheless, attempts are being made to construct specialized languages. Thus, Evert, one of the authors of the work [38] mentioned in the foregoing, developed CELLDYN [35]- a digital program for modelling the dynamics of cell populations. That program (language) with a fixed functional scheme of cell cycle makes it possible, by varying quantitative indices, to study the kinetics of population growth and the distribution of cells among cycle phases. The shortcoming of the program which confines its area of application is the fixed character of the functional scheme. Its advantage is great ease of use for non-professional programmers.

Described in [9] is another computer simulation program called CELLSIM. That program can be rightly named a simulation language enabling construction of much more complicated models. Languages typified by CELLDYN and CELLSIM may be used effectively in solving a fairly wide, though still limited, range of cell biology problems. Their potentialities are much less extensive than those of the general-purpose languages for simulating discrete systems.

Given below by way of illustration are programs in CELLSIM (example from [9]) and GPSS/360 realizing a growing population model in which the cells proceed in succession through all the four phases of the mitotic cycle (G_1,S,G_2,M) and complete it by undergoing division, both daughter cells reentering the cycle (Figure 1.4).

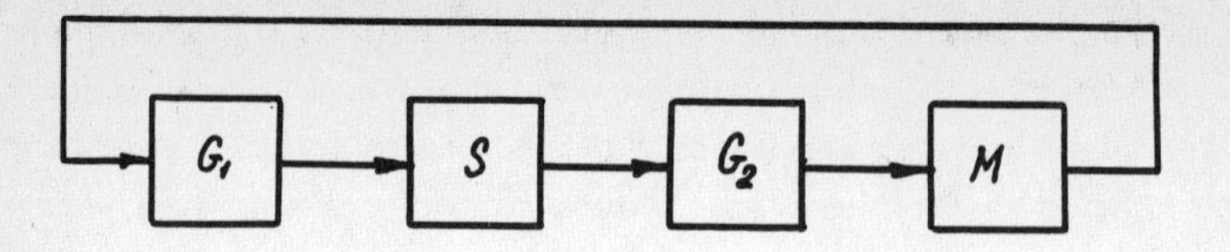

Figure 1.4. Functional block diagram of a growing cell population.

```
CELL TYPES   TYPA;
STATES (1)   G1,S,G2,M;
FLOW (1)   G1-S(ALL), S-G2(ALL), G2-M(ALL); M-G1(ALL);
TIME IN STATES(1) G1: NORMAL(5,1), S: NORMAL(10,2);
                  G2: NORMAL(3,6); M: NORMAL(2,4);
PROLIFERATION(1) M:2;
INCOLUM(1)   1000;
REPORTS  1.0 HOURS;
GRAPH TOTAL S(1)/TOTAL 100;
SIMULATE   50.0 HOURS;
```

This program is CELLSIM is easily understood. All the cells pass from one phase to another, all of them divide in mitosis into two daughter cells which reenter the cycle. Cycle phase durations are normally distributed random variables, their distribution parameters being prescribed for each cycle phase. In this instance simulation time is equal to 50 hours, data output takes place every hour of the model time.

The same model realized in GPSS/360 has the following appearance:

```
        SIMULATE
1       GENERATE  ,,,1
2 CYCL1 ASSIGN    1,4
3 CYCL2 ADVANCE   V1
4       LOOP      1,CYCL2
5       SPLIT     1,CYCL1
6       TRANSFER  ,CYCL1
7       GENERATE  ,,,500
8       TERMINATE 1
```

Here no means for data output or prescribed distributions for cycle phase durations are indicated. However, they are quite simple and present no difficulties in constructing GPSS models.

Operator (block) one enters one cell into the model at the start of simulation. In operator two, written in parameter 1 for each cell is the number of mitotic cycle phases that the cell will pass through. Operator four decreases that number by 1 after completion of each phase until it is equal to 0. A cell thus proceeds in succession through the four phases of the cycle. In the meantime, the values of parameters (mean duration and variance) corresponding to different phases are chosen according to the value of parameter 1 (G_1-4, S-3, G_2-2, M-1) and with variable V1 a pseudorandom number is generated with distribution parameters fixed for the phase concerned. Block three (ADVANCE) delays the cells for the time V1 (simulates transit through the cycle phase), blocks five and six of the model realize the process of cell division (SPLIT) at the end of phase M into two daughter cells which pass (TRANSFER) into the G_1 phase.

Blocks seven and eight prescribe the end of simulation time - 50 hours. Time in this GPSS program is discrete, the minimum step being 0.1 hour.

From the viewpoint of simplicity the above example demonstrates a certain advantage of CELLSIM for simulating a simple system.It is much easier for a user to acquire the skills of employing that language than GPSS. However, the models presented in Chapters II and IV of the present study cannot be realized by means of CELLSIM and in that case it pays to master GPSS.

Our own specific experience prompted us to use the GPSS/360 simulation language, its potentialities being practically in full accord with the aims in view. Among the important virtues of GPSS is the possibility of modelling complex unconventional schemes of cell systems and a comparatively large variety of means for statistical inference from the results of model trials. Its disadvantages are the absence of service programs for statistical analysis of non-stationary stochastic process realizations and of software for planning simulation model trials. A reason not least important for our choosing GPSS was the availability of excellent manuals on that language [6, 18, 55] . In the opinion of many specialists in computer facilities Schriber's manual [5] is one of the best handbooks on programming.

In the context of the foregoing discussion it seems appropriate to touch on the prospect for further development of simulation software aimed at solving biological problems. In so doing it should be noted that a specific language would be required to tackle problems of cell kinetics and a different one to simulate physiologic systems, etc. At the same time a researcher can always choose from the existing plentiful software a universal language best suited to the problem concerned.

Directly following from the foregoing is our view that there is little point in developing "biological" simulation languages.

It should not be neglected, however, that as the appearance of software for simulating computers is largely due to their extensive application, the development of biological languages will be well justified when computer modelling accompanies practically any real biological experiment.

Moreover, the availability of less intricate languages and packages of appied programs (easier to study and to use but not of lesser potentialities) may prove a stimulating factor in the development of biology into an exact science.

Therefore, in our opinion, the question as to whether there is need for further specialization of modelling systems should be answered in the affirmative.

1.6. Possible Uses of Simulation Modelling

A simulation model of a complex biologic system is invariably multiparameter. Reliable identification of certain unobservable parameters of a model is feasible only when a complex of experimental observations provides information predominantly on those parameters, and it is proved, in the meantime, that the values of the other parameters of the model are of little consequence for the identification procedure. Thus, in Chapter III, applying a simulation model to the analysis of experimental evidence of radiation effects on a cell culture, we ascertain, in a special set of simulation trials, that survival of irradiated cells depends primarily on radiosensitivity averaged over the mitotic cycle (rather than on the type of its distribution over the cycle), which permits reducing substantially the number of free parameters in reproducing cell recovery from radiation damage.

The simulation experiments described in Chapter III involved identification of parameters of a model discussed in Chapter II by minimizing the sum of squared deviations of simulation results from the data of a real radiobiological experiment.

Confidence in the right choice of a model and in correct evaluation of its parameters may be enhanced whenever the following two-stage investigation proves realizable. In stage one parametric identification of the model is carried out by a wide complex of different experimental findings (let us call it complex A). In stage two, at fixed parameters, model predictions and real observations comprising another, idenpendent complex of experimental situations (complex B), are tested for agreement. Such possibility of transition from reproducing on a model complex A data to reproducing complex B without additional adjustment of parameters in practice presents itself extremely rarely, but its realization should always be sought since it is important to assess the "predictive power" of a model. It may be that transition from complex A to complex B will require additional evaluation of a single model parameter. Such a situation, considered in Chapters III and IV, should also be regarded as fortunate for the investigator, especially if after estimating the parameter, it is still possible to finally verify the model by some experimental complex B data not used previously for parametric identification.

Identification of unobservable parameters of a simulation model is an indispensable and very important step in applied investigation. However, on no account should it be regarded as the principal, much less the sole aim of such an investigation. Let us now list the possible uses of the models discussed in the present study.

1. Verification of the validity of present-day concepts of the temporal organization of cell renewal processes. General concepts of the cell cycle structure, particular interactions between individual subpopulations and mechanisms regulating cell reproduction and differentiation evolved as a result of generalization and analysis of a great body of indirect experimental evidence. Such analysis invariably relies on the use of different, at times greatly simplified, mathematical models for individual phenomena of cell kinetics. Thus, even at this stage there is a possibility of erroneous interpretation of experimental observations whose consequences may manifest themselves in the reconstruction of the general pattern of cell renewal organization in a particular tissue. Nor is there a quarantee against possible errors in the next stage of the research, i.e. integration of separate results into a harmonious system of concepts of present-day cell biology. The point is that in that stage one has to accomplish synthesis of a very complex model and ascertain that the model explains the entire diverse experience gained by investigators in that field. At this stage simulation modelling appears to be a realiable tool of research and a fairly simple one compared to the development of ma-

thematical theory. Reproduction in simulation experiments of a wide range of actually observable phenomena strengthens confidence in the validity of the existing views on the organization of cell life cycle and facilitates detection of sources responsible for any erroneous interpretations or contradictions. The foregoing also applies fully to investigation into the pattern of damaging effect on renewing cell populations.

2. Formulation of new hypotheses of proliferation and differentiation processes involving intact and damaged cells. What is implied are natural science hypotheses which prompt the investigator to stage new biologic experiments or to reappraise conclusions drawn from traditional interpretation of experimental evidence. Here are some of such examples considered in the subsequent Chapters of the book. Discussed in Section 3.4 is the effect of higher survival rate for irradiated cells resulting from delay in their subculture for testing the clonogenic potentiality. That effect is usually interpreted as manifesting a phenomenon of potentially lethal radiation repair. In so doing, it is implied that no increase in the survival rate for irradiated cells is an indication that no repair of cells from any damage that limits their reproductive ability is taking place. Results of simulation experiments, however, show the existence of conditions (the true state of exponential cell population growth) under which the effect of higher survival may be lacking, despite repair processes actively proceeding in the damaged cells. These findings prompted us to suggest [19] the presence of resting cells in the exponentially growing cell culture LICH which was confirmed independently by analysis of radioautographic studies of the culture.

Another conclusion drawn from the results of stimulation modelling and worthy of further experimental (and, possibly clinical) investigation is also cited in Section 3.4. It can be formulated as follows: resting cells may simultaneously exhibit both higher sensitivity to radiation damage and greater ability for postradiation repair than proliferating cells of the same line. Another example is the marked differences in the shapes of survival curves for twice irradiated cells of radiocurable and radioincurable human tumour explants. Simulation modelling may account for these differences by the dissimilar character of response to repeated irradiation from the system of intracellular repair of radiation damage. This explanation, in turn, suggests better protection against irradiation of the system of repair of radioincurable tumour cells. Finally, the simulation model considered in 4.2 for the "crypt-villus" system of the small intestine epithelium in mice predicts that, under certain conditions of irradiation, futher inhibition of intracellular postirradiation repair processes may result in alleviating the intestinal syndrome of radiation disease. This seemingly paradoxical prediction was borne out in a direct radiobiologic experiment.

Indeed, the most perturbing and, not infrequently, the most conducive moment for one developing a simulation model is when its prediction comes into conflict with biological observations. It is at that moment that the fate of the model is at stake. But that moment may also lead to discovery of hitherto unknown aspects of the phenomenon under study or even of new principles of prime biological importance. Revision of the functional scheme of a model or its elements necessitated by a discrepance between theory and experiment enriches considerably the knowledge of the object under simulation and sometimes may prompt a constructive program for its further experimental study.

3. Substantiation of methods for the mathematical modelling of cell systems and appraisal of their mathematical (analytical) limits of applicability. Two problems of that type will be dealt with in Chapter V. One is concerned with the accuracy of evaluating the so-called q-index which characterizes the cell flow into a given cycle phase and which is calculated from data on the temporal pattern of a number (or fraction) of cells in that phase. In some particular cases the variance of this indirect indicator of cell kinetics may be obtained by mathematical tools of the theory of branching processes. In simulation experiments considerably more extensive studies can be conducted [80] which would give evidence not only of the accuracy of q-index estimators for different states of cell kinetics but also of the robustness of the estimator in relation of violations of certain prerequisites of the model. The other problem is associated with ascertaining the limits of applicability of conventional methods for estimating the temporal parameters (mean duration and duration variance) of mitotic cycle phases by labelled mitoses curves. Such methods are quite appropriately applied to populations in a stationary (strict-sense) state or in that of steady exponential growth. Simulation modelling has shown that transient processes occurring in cell kinetics may significantly distort the shape of the labelled mitoses curve. In other words, the use of conventional methods for analysis of the labelled mitoses curve, with cell flows through the mitotic cycle changing irregularly with time, may lead to errors in estimating temporal parameters of the cycle. These findings cast doubt on the assu-

mption [32] that regeneration of the small intestine epithelium in the mammals after acute radiation damage is characterized by an appreciable decrease in the mean duration of the mitotic cycle of enterocytes. Arriving at such results by purely mathematical means would require a much greater effort. Needless to say the problems considered in the book do not exhaust the potentialities of simulation modelling as a tool of research promoting the development of the mathematical theory of cell systems. No doubt, the converse is also true: an analytical model whose properties are well understood can be used to validate a simulation.

In recent years analytico-statistical methods for describing and studying complex systems and processes have been intensively developing in various engineering fields. They mirror the attempt to make the most of mathematical apparatus for solving specific problems and to resort to computer-aided simulation only in case of insurmountable analytical difficulties. Unfortunately, we do not know of any studies employing that combined approach to problems of cell biology.

Simulation modelling may be also of advantage in statistical analysis of some quantities observed in a biologic experiment, e.g. in verifying statistical hypotheses of certain parameter values for the system involved when there are no indications of the sample distribution of the parameter estimators. Indeed, information on that distribution obtained from simulation trials may be used for validating the choice of a parametric family of distributions. The foregoing, of course, applies to the intrinsic (unobservable) parameters of a model.

4. Inquiry into factors governing cell response to irradiation and to other damaging agents. What is implied here is, in effect, investigation into factors governing the response of a cell population model to modelled radiation. As for a similar real study, however, for most cases there are no adequate experimental techniques.

In solving this class of problems, effective use may be made of the methods of mathematical theory of experimental design and of factor variance analysis. An example of applying this approach to exploration of a relative contribution of some cell system characteristics to the resultant effect of radiation is discussed in Section 3.6

5. Search for general principles of optimal dynamic regime of irradiation (or chemotherapheutic treatment) of tumour cell populations to maximize therapeutic effect. We are concerned solely with the general principles inasmuch as the extremely high variability of tumour and host characteristics is a serious obstacle to individualizing search on a model for optimal therapy protocols. The attempts to solve

this problem reported in the literature [13, 62] are based on comparatively simple models of cell population response to external action and on methods of optimal control theory. The use for this purpose of simulation modelling could make for more realistic descriptions of cell population dynamics in normal and tumour tissues. In conducting such investigations, it is expedient to integrate a complex simulation model with software for stochastic optimization methods. In so doing optimization problems are considered to be parametric, the definition of optimum control problems presenting, in that case, considerable difficulties. Application of a digital stochastic simulation model for the kinetics of irradiated cells [79] has shown that, from the standpoint of maximum inhibition of cell population growth, there exists an optimum time interval between two irradiations of equal magnitude. However, the simulations being long and complicated, that approach has gained no wide acceptance. It is hoped that the modern super-computers now at the disposal of many laboratories will help to overcome these difficulties in the immediate future.

6. Investigation of general regularities in the action of modifiers of radiation (and other types of) damage to cells. The mechanism of action of certain known radiomodifiers can be formalized and, in that case, simulation modelling will facilitate preliminary review of possible effects of combined action. A case in point is the above-mentioned prediction by a model of an unexpected effect of a repair inhibitor on the irradiated small intestine epithelium. In this field, too, simulation may be expected to yield interesting new results.

The above-enumerated ways of utilizing simulation modelling will be illustrated by concrete examples in the subsequent sections of the book. Those examples are preceded by a comprehensive description (Chapter II) of the model for radiation effect on cell culture and of the conditions for studying its properties in different experimental situations. The model of organized tissue (Chapter IV) is described in less detail since many of its elements are identical to those of the model considered in Chapter II.

REFERENCES

1. Absher, R.G. and Absher, P.M. Mathematical models and computer simulation of proliferation of human diploid fibroblast clones, J.Theor.Biol., 72, 627-638, 1978.
2. Arndt, S. and Langer, O.U. DIGSIM - ein Programm zur digitalen Simulation und Optimierung verfahrenstechnischer Systeme, Chem.

Techn., 24, 2-12, 1972.
3. Aroesty, J., Linkoln, T. and Shapiro N. Tumor growth and chemotherapy: Mathematical methods, computer simulation and experimental foundation, Mathem, Biosci, 17, 243-300, 1973.
4. Bharucha-Reid, A.T. Elements of the theory of Markov processes and their applications, MC Graw-Hill, New York, 1960.
5. Blumenson, L.E. A comprehensive modeling procedure for the human granulopoietic system, Mathem. Biosci., 26, 217-239, 1975.
6. Bobillier, P.A., Kahan, B.C. and Probst, A.R. Simulation with GPSS and GPSS V, Prentice-Hall, Inc., Englewood Cliffs, New Jersey, 1976.
7. Creekmore, S.P., Aroesty, J., Willis, K.L., Morrison, P.F. and Lincoln, T.L. A cell kinetic model which includes heredity, differentiation, and regulatory control, In; Biomathematics and cell kinetics, Elsevier / North-Holland Biomed. Press, 255-267, 1978.
8. Dahl. O.-J. Discrete event simulation languages, Norsk Regnesentralen, 1966.
9. Donaghey, C.E., Drevinko, B. A computer simulation program for the study of cellular growth kinetics, Comput. and Biomed. Res., 8, 118-128, 1975.
10. Düchting,W., Denl, G. Spatial structure of tumour growth: a simulation study, IEEE transactions on systems, man, and cybernetics, SMC-10, 292-296, 1980.
11. Düchting, W., Vogelsaenger, Th. Three-dimentional pattern generation applied to spheroidal tumor growth in a nutrient medium, Intern. J. Biomed. Comput., 12, 377-392, 1981.
12. Düchting, W., Vogelsaenger, Th. Über Fortschritte auf dem Gebiet der Simulasion von Tumorbehandlungen, Regelungstecknik, 31, 3--8, 1983.
13. Eisen, M. Mathematical models in cell biology and cancer chemotherapy, Springer-Verlag, Berlin, Heidelberg, New York, 1979.
14. Eisenfeld, J., Relation between stochastic and differential model of compartmental system, Mathem. Biosci., 43, 289-305, 1979.
15. Evert, C.F. CELLDYN - digital programm for modeling the dynamics of cell, Simulation, 2, 55-60, 1975.
16. Feller, W. Die Grundlagen der volterraschen Theorie des Kampfes und Desein in Wohrscheinlisheits-theoretischer Behandlung, Acta biotheor., 5, 11-40, 1939.
17. Gopolsamy, K. Dynamics of maturing population and their asymptotic behaviour, J. Mathem. Biol., 5, 383-398, 1978.
18. Gordon, G., System simulation, Prentice-Hall, Inc., Englewood, Cliffs, New Jersey, 1978.
19. Gushchin, V.A., Zorin, A.V., Stephanenko, F.A. and Yakovlev, A.Yu. On the interpretation of the recovery of cells from potentially lethal radiation damage in stationary cell culture, Studia Biophys., 107, 195-203, 1985 (In Russian)
20. Hagander, P. Simulation and flow-systems analysis of the cell cycle. Mathem. Biosci, 48, 241-265, 1980.
21. Harris, T.E. The theory of branching processes, Springer-Verlag, Berlin, 1963.
22. Hauser, F. and Nečas, E. Use of a model for investigation of hypotheses about proliferation control in a stem cell population, In: Simulation of systems in biology and medicine, Dum Techniky, Praha, 202-210, 1982.
23. Hoffman, J.G., Metropolis, N. and Gardiner, V. Digital computer study of cell multiplication using Monte-Carlo methods, J. Nat. Cancer Inst., 17, 175-188, 1956.

24. Honda, H., Toshiteri, M. and Tanabe, A. Establishment of epidermal cell columns in mammalian skin: Computer simulation, J. Theor. Biol., 81, 71-80, 1979.
25. Jagers, P. Branching processes with biological applications, John Wiley and Sons, N.Y.-Sidney-Toronto, 1975.
26. Jansson, B. Simulation of cell-cycle kinetics based on a multicompartmental model, Simulation, 10, 99-108, 1975.
27. Kendall, D.G. On the generalized birth-and-death process, Ann. Math. Stat., 19, 1-15, 1948.
28. Kimmel, M. Cellular population dynamics, Mathem. Biosci, 48, 225--239, 1980.
29. Kindler, E. Simulating languages, Sntl-Nakladatelstvi technike literatury, Praha, 1980 (In Czech).
30. King-Smith, E.A. and Morley, A. Computer simulation of granulopoiesis; normal and impaired granulopoiesis, Blood, 36, 254-259, 1970.
31. Lamb, J.R. A regenerating computer model of the thymus, Comput. and Biomed, Res., 8, 379-392, 1975.
32. Lesher, S., Cooper, J., Hagemann, R. and Lesher, J. Proliferative patterns in the mouse jejunal epithelium after fractionated abdominal X-irradiation, Current Topics in Radiat. Res., 10, 229-261, 1975.
33. Lincoln, T., Morrison, P., Aroesty, J. and Carter, G. Computer simulation of leukemia therapy, Cancer treatment r Rep. 60, 723-739, 1976.
34. Loefler, T. and Wichmann, H. A comprehensive mathematical model of stem cell proliferation, Cell Tissue Kinet., 13, 543-561, 1980.
35. Mackey, M.C. Some models in hemapoiesis: predictions and problems, Biomathematics and Cell Kinetics, 8, 23-38, 1981.
36. Malinin, A.M. and Yakovlev A.Yu. The fraction of labelled mitoses curve in different states of cell proliferation kinetics. III. Modelling experiments, Cytology, 18, 1464-1469, 1976, (In Russian).
37. Mary, J.Y. A model of granulopoiesis in normal man, In: Biomathematics and Cell Kinetics, Biomedical Press, Amsterdam New York Oxford, 269-285, 1978.
38. Mauer, M.A. Evert, C.F., Lampkin, B.C. and McWilliams, N.B. Cell kinetics in human acute limfoblastic leukimia: Computer simulation with discrete modeling techniques, Blood, 41, 141-154, 1973.39.
39. Mode, C.J. Multitype age-dependent branching processes and cell cycle analysis, Mathem. Biosci., 10, 177-190, 1971.
40. Moiseev, N.N. (editor). The present-day status of the theory of operation studies, Nauka, Moscow, 1979 (In Russian).
41. Monichev, A.Ya. A quantitative model of normal granulopoiesis in man, Dep. VINITI, N 483-80, Moscow, 1980 (In Russian).
42. Nečas, E., Hauser, F. and Neuwirt, J. Computer model of hemopoietic stem cell population testing a possible role of DNA synthesizing cell in proliferation control, Blood, 41, 335-336, 1980.
43. Oldfield, D.C. The discontinuity equation for cellular population, Bull. Mathem. Biophys., 28, 545-554, 1964.
44. Pertsev, N.V. and Marchuk, G.I. Mathematical models of hemopoiesis, In: Mathematical Modeling in Immunology and Medicine, North-Holland Publishing Company, 197-202, 1983.
45. Poluektov, R.A. (editor). The dynamic theory of biologic populations, Nauka, Moscow, 1974 (In Russian).
46. Ravkin, J.A. and Pryanishnikov, V.A. Simulation of the cell cycle of complex populations, Cytology, 19, 625-631, 1977 (In Russian).

47. Rotenberg, M. Theory of distributed quiscent state in the cell cycle. J. Theor. Biol., 96, 495-509, 1982.
48. Rubinov, S.I. A maturity-time representation for cell population, Biophys. J., 8, 1055-1073, 1968.
49. Rubinov, S.I. and Lebowitz, J.L. A mathematical model of neutrophil production and control in normal man, J. Mathem. Biol., 1, 187-225, 1975.
50. Saito, N. Asymptotic regular pattern of epidermal cell in mammalian skin, J. Theor, Biol., 95, 591-593, 1982.
51. Sakharov, M.P. and Serebrovsky, A.P. Mathematical modelling of malignant cell populations, IAT, Moscow, 1975 (In Russian).
52. Scherbaum, O. and Rasch, G. Cell size distribution and single cell growth in Tetrahymena Pyriformis CL. Acta Pathol. Microbiol. Scand., 41, 161-182, 1957.
53. Schotz,W.E. Continuous labeling indices: CLI(t),and CLM(t), J. Theor. Biol., 34, 29-46, 1972.
54. Schotz, W.E. Double label estimation of the mean duration of the S-phase, J. Theor. Biol, 46, 353-368, 1974.
55. Schriber, T.J. Simulation using GPSS, John Wiley and Sons, New York, London, Sydney, Toronto, 1974.
56. Schwarze, G., Simulation-kontinuierliche Systeme, Technik, Berlin, 1976.
57. Shackney, S.E. A computer model for tumor growth and chemotherapy, Cancer Chemother. Rep., 54, part 1, 399-429, 1970.
58. Shannon, R.E. Systems simulation. The art and science, Prentice--Hall, Inc., Englewood Cliffs, New Jersy, 1975.
59. Shin, K.G. and Pado, R. Design of optimal cancer chemotherapy using a continuous time-state model of cell kinetics, Mathem. Biosci., 59, 225-248, 1982.
60. Shvytov, J.A. Mathematical models of the growth of population numbers, In: Mathematical Modelling in Biology, Nauka, Moscow, 1975 (In Russian).
61. Skipper, H.E. Improvement of the model systems, Cancer Res., 29, 2329-2333, 1969.
62. Swan, G.W. Optimization of human cancer radiotherapy, Springer--Verlag, Berlin, Heidelberg, New York, 1981.
63. Takahashi, M. Theoretical basis for cell cycle analysis, J. Theor, Biol., 18, 195-209, 1968.
64. Thames, H.D. Mathematical model of dose and cell cycle effects in multifraction radiotherapy. In: Lecture Notes in Pure and Appl. Mathem, 58, 51-105, 1980.
65, Thames, H.D. and White, R.A. State-vector models of the cell cycle. I. Parametrization and fits to labeled mitoses data, J. Theor. Biol., 67, 733-756, 1977.
66. Theodoroff, T.J. and Olsztyn, J.T. DYANA, dynamics analyzer-programmer, Proc. EJCC, 1958.
67. Toivonen, H. Transient cell kinetics, Frenchellin Kirjapaino Oy, Helsinki, 1980.
68. Toivonen, H. and Rytömaa, T. Monte Carlo simulation of malignant growth, J. Theor. Biol, 78, 257-267, 1978.
69. Valleron, A.-J. and Frindel, E. Computer simulation of growing cell populations, Cell Tissue Kinet., 6, 69-79, 1973.
70. Valleron, H.-J. and Macdonald, P.D.M. (editors). Biomathematics and Cell Kinetics, Biomedical Press, Amsterdam-New York-Oxford, 1978.
71. Von Foester, J. Some remarks on chaging population. In:The Kinetics of Cell Proliferation, Grune and Stratton, New York, 382--407, 1959.
72. Wasan, M.T. Stochastic approximation, University Press, Cambridge , 1969.
73. White, R.A. The use of population projection matrices in cell kinetics, J. Theor.Biol., 74, 4967-4977, 1978.

74. Yakovlev, A.Yu. and Yanev, N.M. The dynamics of induced cell proliferation within the model of a branching stochastic process. I. Numbers of cells in successive generations, Cytology, 22, 945-953, 1980 (In Russian).
75. Yanev, N.M. and Yakovlev, A.Yu. The dynamics of induced cell proliferation within the model of a branching stochastic process. II. Some characteristics of the cell cycle temporal organization, Cytology, 25, 818-825, 1983 (In Russian).
77. Yanev, N.M. and Yakovlev, A.Yu. On the distribution of marks over a proliferating cell population obeying the Bellman-Harris branching process, Mathem. Biosci., 75, 159-173, 1985.
77. Yakovlev, A.Yu. and Zorin, A.V. On the simulation of reliability of renewal cell systems, Cytology, 24, 110-113, 1982 (In Russian).
78. Yakovlev, E.I. Computer-based simulation, Nauka, Moscow, 1975 (In Russian).
79. Zorin A.V., Gushchin, V.A., Stefanenko, F.A., Cherepanova, O.N. and Yakovlev, A.Yu. Computer simulation of kinetics of irradiated cell populations in tumours, Experimental Oncology, 5, 27-30, 1983, (In Russian).
80. Zorin, A.V. and Yakovlev, A.Yu. The properties of cell kinetics indicators. A computer simulation study, Biom. J., 28, 347--362, 1986.

II. A SIMULATION MODEL FOR IN VITRO KINETICS OF NORMAL AND IRRADIATED CELLS

2.1. Introduction

The model proposed here simulates development in time of a cell population which, depending on a chosen model structure and initial values of the parameters, may correspond to either an exponentially growing or stationary culture of normal or tumour cells, as well as to a stem-cell population from embryonal or somatic definitive renewing tissues. It is designed primarily for studying the effect of radiation on the kinetics of cell populations following a single or fractionated exposure. Radiation effect is assessed from the clonogenic capacity of cells. The computer realization of the model was accomplished with the aid of the GPSS/360 language. The individual blocks of the model will be described and substantiated from the standpoint of present-day radiobiological concepts.

2.2. Basic Premises of Unirradiated Cell Population Model

In accordance with the schematic representation shown in Figure 2.1 a cell may be in one of the two states: in the resting state (G_0) or in that of the mitotic cycle (MC). Daughter cells directly after their birth (termination of the M phase) pass into the resting phase. Transition of cells from G_0 to the G_1 phase is regarded as probabilistic, depending upon the total number of cells within the system and upon the ratio of cells in the G_0 state to those in the MC. For the purpose of describing in detail the cells' entry into the MC let us consider the possible procedures for introducing cells into the model.

Procedure 1. At the beginning of simulation a certain number of cells N_c is specified which will be maintained throughout the entire period of simulation or until the exposure to some effect (e.g.,irradiation). At once one cell is introduced which, as a result of subsequent divisions, will bring the population numbers to N_c. Thus, a desynchronized progress of the cells through the cycle is induced in a "natural" way.

Procedure 2. The specified number of cells is introduced at a fixed moment (ordinarily, at the beginning of simulation), two possible goals being attained: the construction of synchronized and asynchronous populations. The cells may be introduced into one or several of

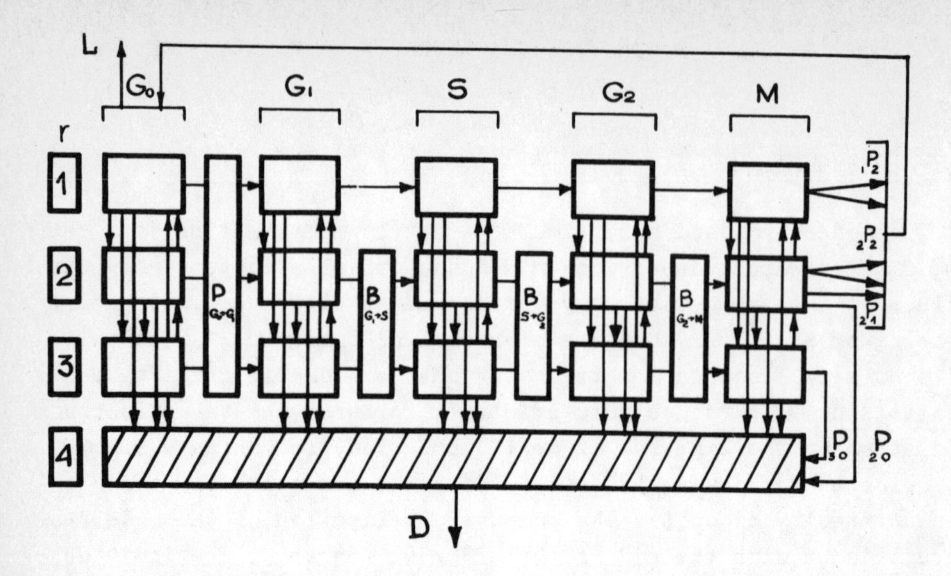

Figure 2.1. Functional block diagram for the model of an irradiated cell system
r=(1,2,3,4)-levels of cell damage. Transition to r=4 level is irreversible and results in cell death (D)
$P_{G_0-G_1}$ - probability for a resting cell to pass to the MC
B_{i-j} - irradiation caused delays in the progress of cells through the MC
$_rP_k$ - probability for the rth level cell to produce k descendants (for detais see text).

the five phases of the life cycle - G_1, S, G_2, M, and G_0 - in two ways: either at the onset of some phase - synchronized population, or by distribution over the phase (over phases, over the entire MC) - asynchronous population. The rate of entry of the G_0 cells into the MC is determined by the probability $P_{G_0-G_1}$ which is specified graphically (Figure 2.2).

The transition of cells from the G_0 state to the MC may occur either at the moment of some changes within the cell system (such as termination of mitosis, movement to the G_0 state of one or several cells, cell death) or at the moment of irradiation. If the population has reached the pre-set N_c level in cell numbers, all the "superfluous " cells are eliminated from the model. In so doing, the G_0 cells alone are eliminated, the choice of cells which are removed being made in a random manner, i.e. elimination may affect a cell which

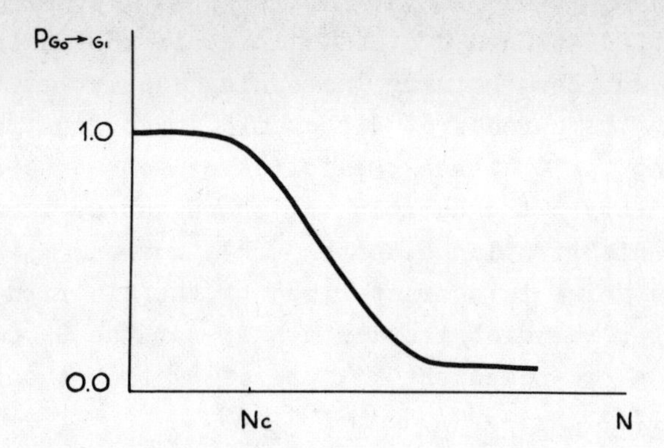

Figure 2.2. Relationship between the probability of a resting cell transition to the MC and total population numbers (explanation in the text).

has just entered the G_0 state or, by equal chance, one that has resided in it for some time.

For transition to the MC cells are likewise chosen in a random manner. For each G_0-cell a random number is generated which is distributed uniformly within the interval $[0,1)$. The sampled value of this number is then compared to the probability $P_{G_0-G_1}$ depending on the argument $N_{G_0} + 2N_{MC}$. Thus, cell death induces proliferation which gains in scope with the increasing number of cell deaths, and with the growing numbers of the whole population the proliferative activity of cells diminishes. There appears to be no fundamental difficulty in introducing into the model the relationship between the probability of cell transition to the cycle and the local density of cell population, rather than the population numbers. However, the purposes pursued by this study do not call for such complication of the computer program. Cells embarking upon mitotic division traverse subsequently and independently of each other the phases G_1, S, G_2 and M. The durations of the cycle phases are considered as independent equally distributed random variables. In this model truncated on the left normal distribution of each variable is used. Thus, the durations of cycle phases differ from each other only in numerical parameters: mean values $\bar{\tau}_i$ and variances σ_i^2. Experience shows that the utilization for the analysis of cell kinetics of different standard unimodal distributions of the MC phase durations yields practically the same results [32]. The model provides the possibility of describing the ramified structure of mitotic cycle phases the reality of whose existence has been gaining increasing support in the past few years [27] . For the purpose of dichotomic stratification of the flow of cells entering the G_2 phase the following scheme is considered : with the probability p the duration of the G_2 phase is a random variable with the distribution density $f_1(x)$, whereas with the probability (1-p) its value is characterized by the distribution density $f_2(x)$. Apparently, the distribution density of the G_2 phase duration under review is a two-component mixture of distributions :

$$f(x) = pf_1(x) + (1-p)f_2(x).$$

Equal (normal) families of distributions $f_1(x)$ and $f_2(x)$ were chosen; thus, the "short" G_2 phase differed from the "long" G_2 phase only by numerical parameters.

Upon completion of the mitotic division of an unirradiated cell two daughter cells appear with the probability 1 which immediately pass into the G_0 phase. Thus, the temporal organization of the cell MC described above corresponds to the model which, though developed by many authors [6, 7, 10, 11, 68, 69] came to be known in the literature under the name of the Smith and Martin model. Some studies demonstrated a satisfactory agreement between that model and the data on the distribution of the mitotic cycle duration obtained by time--lapse cinematography of cell cultures [63, 64, 68, 69] . Other authors on the basis of similar observations raised objections to the Smith and Martin model [25, 26] . The discussion that developed in the literature [25, 26, 50, 68] focussed mainly on the interpretation of the evidence of time-lapse cinematography and was eventually of a purely technical nature. Some specific features of the kinetics of induced cell proliferation, revealed in the analysis of the rat liver regeneration [85] , are also at variance with the predictions of the Smith and Martin model. The criticism, however, is aimed at the proposition of the Smith and Martin model in its initial form according to which all of the daughter cells which are formed after division pass invariably into the G_0 state, while those residing in the G_0 phase have equal chances of being moved to the mitotic cycle under the action of proliferative stimulus.

Svetina and Žeks [70] have extended the basic Smith and Martin model on the assumption that the rate constant for the transition from the G_0 state to the MC depends on the time a particular cell has already spent in the G_0 state.

We believe that at the present stage of the development of radiobiology further elucidation of the particulars of the resting state structure and of the regularities inherent in the transition of daughter cells into the next mitotic cycle would be premature and unessential for the simulation of many radiation effects which are evaluated by the criterion of survival (clonogenic capacity) of irradiated cells. It was for this very reason that the initial version of the Smith and Martin model was used as the basis for the functional diagram presented in Figure 2.1. Since the simulation model proposed here is oriented at in vitro studies of cell systems, it includes no description of the processes of cell maturation and differentiation. The heterogeneity of a cell population may be accounted for in a model by the introduction of cell subpopulations C_k corresponding, for instance, to zones of different monolayer densities. Subpopulations C_k differ from each other in such properties as are of interest to a researcher: radiosensitivity, radiation damage repair, the scope of processes of spontaneous cell death, etc. The interaction between subpopulations C_k is ensured by unidirectional transition of cells from C_{k-1} to C_k. A more detailed description of the principle of simulating the heterogeneity of a cell system is given in [88] .

2.3. Basic Principles of Simulating Cell Inactivation after Exposure to Ionizing Radiation

Radiobiologic effects observed under in vitro conditions with exposure to acute irradiation are usually interpreted by studying the shape of the effect-dose curves which characterize changes in the survival (clonogenic capacity) of cells due to increasing doses of irradiation. Henceforth, the survival-dose relationship normalized for survival in unirradiated control will be designated as S(D). For a long span of time the methods employed for the analysis of such relationships were intimately associated with the principles of the target theory. Among the difficulties encountered in the formal application of that theory to the analysis of the S(D) curves, obtained in numerous experiments with mammalian cells, those most commonly discussed in the literature are:

(1) the need of describing the shouldered S(D) curves at low irradiation dose values;
(2) the need of including into the model the nonzero initial slope of the S(D) curve.

With a view to overcoming the first difficulty many models were proposed which explained the presence of a shoulder in the S(D) curve either by accumulation or interaction of radiation damage of the cell. To describe the S(D) curves obtained by them on the HeLa cell cultures, Puck and Marcus [59] used the equation

$$S(D) = 1 - \left[(1 - e^{-D/D_0})^n\right]^m , \qquad (1)$$

which is known in the literature under the name of a "multihit-multitarget model". In the analysis of data from experiments on mammalian cells (including those by Puck and Marcus [59] with the HeLa culture) it is very often found that m=1 and that a special case of equation (1) can be used

$$S(D) = 1 - (1 - e^{-D/D_0})^n , \qquad (2)$$

i.e. a "singlehit-multitarget model" [40] . The biological essence of models (1) and (2) is such that a certain accumulation of individual radiation (sublethal) damage is required for the manifestation of the effect (evaluated by cell survival).

Kellerer and Rossi [41] suggested that under the action of high LET radiation, lethal effects are accumulated linearly with the dose, while on exposure to irradiation with low LET the accumulation of lethal damage is proportional to the squared value of the dose. This approach led to the equation

$$S(D) = e^{-\alpha D - \beta D^2} , \qquad (3)$$

which realizes the so-called linear-quadratic model [41, 42] . Chadwick and Leenhouts [8] derived the same equation (3) within the framework of the molecular theory of radiation effect on a cell. They suggested that cell survival is determined by the presence of double-strand breaks in the DNA molecule which may result from a single act of simultaneous energy absorption in two DNA chains or from separate acts of energy deposition in closely spaced sites of different DNA strands. In a later study [44] emphasis is laid on interpreting equation (3) in terms of the double-track mechanism of radiation effect whose characteristic feature is the interaction of damage at a short distance, leading to greater biologic efficiency of the second track as compared to the first. Other models were also proposed, introducing minor modifications into the above prototypes. The second theoretical difficulty associated with the nonzero slope of the S(D) curve was overcome only in the formal manner by including in the model a single-hit component and modifying equation (2) [3, 5, 83] as follows

$$S(D) = e^{-D/D_1} [1 - (1-e^{-D/D_2})^n] \quad . \qquad (4)$$

The models expressed in the form of equations (1), (2), (3) and (4) were subjected to constructive criticism in the works of Alper [2] and Goodhead [22] .

Based on the analysis of extensive experimental evidence on damaging and mutagenic effects of irradiation with different LET, Goodhead demonstrated the essential inadequacy of the above-listed models and adduced experimental facts supporting the following alternative model of radiation effect upon a cell :

1. Radiation damage occurs at short distances from the sites of absorption of small amounts of energy, the predominant role in producing damage being played by the single-track mechanism.

2. The number of injuries is proportional to the irradiation dose and dependent upon the LET of the radiation used.

3. Some damage can be repaired by the intracellular repair system whose efficiency drops with the increasing radiation dose.

Alper [2] also came to the conclusion that radiation damage occurs in accordance with the single-hit mechanism and that the relatively low sensitivity of cells in the event of small doses (shoulder on S(D)) is associated not with the multiplicity of targets or the interaction between separate injuries but with the existence of a certain biochemical mechanism of repair. Alper [2] stresses that, as a rule, as the radiation dose increases the efficiency of the repair mechanism declines, apparently due to the exhaustion of the biochemical factors (repair enzymes) involved in it. The author notes that only in the case of exponential survival curves there is reason to believe that either that mechanism does not function at all or, on the contrary, it maintains equal efficiency over the entire range of doses. According to Alper [2] this idea was first set forth in [58] . Then, Sinclair [67] advanced the hypothesis of the existence of a certain factor Q, fluctuations in whose concentration determine variations in the magnitude of the S(D) curve shoulder as the cells traverse the mitotic cycle. Thus, came into being such terms as the Q factor, Q repair and Q damage [1, 2] . It is disturbances in the Q repair system that may account for the dependence of the S(D) curve shoulder and the ability of cells to repair sublethal and potentially lethal damage on the nutrient composition and the conditions of cultivation [31, 46, 47] . Similarly, the absence of shoulder in the dose-survival curve obtained from early passages of the primary culture of the embryonal human lung cells and its appearance for later culture passages [9] may be explained by temporary loss and subsequent recovery of the Q factor in the intracellular (intranuclear) medium.

Several mathematical models have been proposed to formalize this idea of the existence of a saturable repair system [1, 2, 23, 24, 43, 53]. Those models differ from each other mainly in the analytical form of relationship between repair efficiency and irradiation dose.

The model of Laurie et al. [43] attempts to separate, within the framework of the Q repair concept, the processes of repair of sublethal and potentially lethal damage. For the purpose of describing the stage-wise character of the development of the radiation inactivation of cells, Garret and Payne [20] proposed the Markovian model with continuous time and three discrete states of a target : A - absence of damage, B - presence of reparable damage and C - presence of irreparable damage. The principal element of the model is the concept of "fixation time" which specifies that all reparable damage (B state) becomes irreparable (C state) unless its repair is effected within a certain time interval called "fixation time". Cells suffering greater than the N damage in the C state die the reproductive type of death. Similar features are inherent in the "stochastic" model of Kapult'cevich and Korogodin [38, 39].

Identical approach was also developed for describing the effect of ultraviolet irradiation of cells [21, 33, 34]. According to the Haynes model [33] the S(D) curve shoulder is due to inactivation (lower intensity) of repair processes caused by increasing doses of ultraviolet irradiation, i.e. with the increasing dosage of number of reparable lesions approaches saturation. In his later work Haynes [34] modified the model by introducing the extreme form of relationship between repair efficiency and dose. Haynes' modified model predicts that the absolute number of pyrimidine dimers, excised as a result of excision repair, at first increases with the dose of ultraviolet irradiation, attains its maximum and then diminishes with larger doses.

This type of repair-dose relationship appears quite natural since the radiation effect is an indispensable condition for the induction of repair processes. At the same time, in the case of some cell lines the basal level of repair processes may be rather elevated. Thus, studying HeLa S3 cells, Djordjevic et al. [12] discovered by means of radioautography spontaneous unscheduled DNA synthesis occurring primarily in the G_1 (or G_0) phase of the cell cycle.

Although the said notions (in the version proposed by Haynes) of the regularities of radiation inactivation of cells were adopted by us as the general principle for constructing a model of radiation effect upon pooliferating cells in vitro, they did not take account of many factors which determine the fate of irradiation cells.

The remarkable success of the target theory is due to the fact that it expressed in the quantitative form the contribution of the discrete nature of ionizing radiation and the probabilistic character of radiation energy transfer to cell structures to the ultimate biological effect which characterizes the probability of survival of the cell and its progeny. However, with the accumulation of experimental facts, there was an ever-growing need for a probabilistic description of the response of the cell itself and for an evaluation of the contribution of that response to the radiobiological effect observed at the cell population level [20, 37, 38, 39, 55] . The response of a cell to radiation effect is of stochastic nature. It is determined by a complex of factors which are extremely difficult to take account of within the framework of an analytical model. Thanks to investigations conducted with the aid of time-lapse microfilming, Elkind and Whitmore [17] as far back as 1967 described the general pattern of cell response to ionizing radiation effect. In the years that followed that pattern underwent no substantial modifications, to say nothing of any radical revision. Following the lead of Elking and Whitmore [17] the following classes of cells may be distinguished in an irradiated population:

1. Surviving cells which retain ability for unlimited proliferation (or for a number of divisions sufficient for revealing a clone). The rate of proliferation of such cells may remain unchanged or decrease compared to that prior to irradiation.

2. Cells capable of only a limited number of divisions, precursors of abortive colonies (reproductive death).

3. Cells undergoing lysis immediately or some time after irradiation without division (interphase death).

4. Cells udergoing neither division nor lysis but remaining in the resting state or forming giant cells.

5. Cells whose poogress through the mitotic cycle is delayed and which, after the elimination of the radiation block, may either perish according to the reproductive type or acquire clonogenic capacity as a result of the activities of repair systems.

Thus, the possible outcomes of the mitotic cycle of irradiated cells may be as follows: formation of two normal daughter cells, formation of one sterile and one non-sterile descendants, and formation of two sterile descendants. Each of these outcomes for every individual cell materializes with a certain probability which depends both on the extent of radiation damage and the intensity of processes of its repair. The end result of radiation effect upon a cell is determined by the interaction of two processes: radiation damage and post-irradiation recovery; their formalization is discussed in what follows.

2.4. Process of Formation of Radiation Cell Damage

One of the principal propositions of the model described here is the assumption that in the case of exposure to acute radiation the average time required for final formation (fixation) of radiation lesions is much shorter than the average time interval during which these lesions are either repaired or materialize into the final effect, i.e. cell death. The fate of an irradiated cell is determined primarily by the degree of its damage after exposure to radiation. This may be taken account of by the introduction of "levels of cell damage" characterizing both the number and the biological significance of cell structure lesions resulting from exposure to radiation. Such levels should differ from one another by probabilities of lesion realizations such as reproductive or interphase cell deaths.

Introduced in the proposed model (Figure 2.1) are four levels of damage indexed with the letter "r" (r=1,2,3,4). Corresponding to the level 1 are intact cells, and to the level 4 those irreversibly (lethally) damaged, the two intermediate levels (r=2,3) have been introduced to take into account possibilities of different degrees of cell damage. Thus, the level r=4 represents the absorbing state. The cells in this state are no longer capable of further progress through the MC phases or of radiation damage repair, but for some random (generated by definite probabilistic law) time they remain within the system and are taken account of in the total numbers of the cell population. Transition to the level r=4 is interpreted as interphase cell death.

After irradiation all the cells of the levels r=1,2 and 3 retain the ability to move through the cell cycle phases at the previous rate. Longer durations of MC phases for irradiated cells are modelled by introducing radiation blocks at phase boundaries (see below).

MC outcomes are dissimilar for cells belonging to different levels of damage at the time of completion of mitotic division and are characterized by different random numbers of progeny ν_1, ν_2 and ν_3 for the levels r=1,2 and 3 respectively.

Let us introduce designation ${}_ip_k = \Pr(\nu_i = k)$, $i=1,2,3$; $k=0,1,2$, for the probability that a cell of the ith level gives rise to a given number k of ancestors. The following values of probabilities were chosen in the model under discussion:

level r=1 - ${}_1p_o = 0$, ${}_1p_1=0$, ${}_1p_2=1$;

level r=2 - ${}_2p_o = 1/3$, ${}_2p_1=1/3$, ${}_2p_2=1/3$;

level r=3 - ${}_3p_o = 1$, ${}_3p_1=0$, ${}_3p_2 = 0$

Reproductive cell death may occur with the above probabilities upon completion of each MC. Thus, radiation lesions corresponding to the level r=2 unrepaired in the given MC are inherited in cell generations and may realize in subsequent division cycles.

Adhering formally to the present-day definitions of the notions of sublethal and potentially lethal damage [15, 17, 56, 86] the following interpretation may be offered of the levels of damage: level 1 - no lesions, level 2 - sublethal lesions, level 3 - potentially lethal lesions, level 4 - lethal, irreparable lesions. It should be remembered however, that this is a conventional classification since the notions of sublethal (SLD) and potentially lethal damage (PLD) adopted in radiobiology are not rigorous, being determined through the experimental scheme employed for revealing one or another type of damage. Another reason for combining all known types of damage within the framework of a single mechanism of radiation inactivation of cells is supplied by the conclusion of Elkind et al. [16] to the effect that sublethal and potentially lethal lesions are different sides of one and the same phenomenon. They are similar as regards the type of manifestation in respect of both the direct process of lesion formation and the reverse process of repair. In order to bring the notions of SLD and PLD into a more precise conformity with the levels of cell damage as introduced above, it is necessary to increase the number of such levels which, in turn, complicates the problem of a priori assignment of probabilities ${}_ip_k$.

Thus, as regards the proposed model, the problem of simulating the process of formation of radiation cell damage consists in assigning a certain rule according to which the cells, after irradiation with a specified dose, are distributed among the levels of damage. With a view to formulating such a rule, let us introduce the notion of "elementary dose" (ED) of radiation. Let us apply the term "elementary" to that dose of ionizing radiation on exposure to which a cell of the rth level with the probability P remains at that level, but the one with the probability q=1-P moves to the neighbouring level r+1. Thus, the total dose of irradiation is always a multiple of the elementary dose. The choice of ED value is arbitrary. The lesser ED the smaller may be the interval by which the value of the total dose of irradiation is specified. Depending on the chosen ED value, probability P is selected. In the present study the chosen ED value is equal to 0.25 Gy i.e. the possible values of irradiation doses may be equal to n•0.25 Gy, where n=1,2,...

The procedure for modelling radiation effect is as follows. At the instants of irradiation specified before the beginning of each simulation experiment, probabilities of transition from the levels r=1,2, 3 to the levels r=1,2,3,4 are played for each cell.

Probability P of a cell remaining at the same level of damage and probability (q=1-P) of its transition to a deeper level are equal in the model for all levels. Further the simulation scheme is based on the following considerations.

Let us introduce a random variable X_k which may have values 0 and 1 with probabilities P and q, respectively. Unity corresponds to transition from the level r to the level r+1 with the absorption of one ED (kth from n), and zero signifies that the cell remains at the same level. The value $S_n = X_1 + X_2 + \dots + X_n$ is a sum of random variables.

Introduce the distribution of probabilities

$A_o = Pr(S_n=0)$, $A_1 = Pr(S_n=1)$, $A_2 = Pr(S_n=2)$, $A_3 = Pr(S_n \geq 3)$, which characterizes the effect of total dose n•ED. Let us then introduce the generating function $a(s) = P + qs$. In this case the generating function for the random number S_n has the form $a^n(s)=(P+qs)^n$ and the first three coefficients of $a^n(s)$ will exhaustively determine the required distribution, since,

$$Pr\{S_n \geq 3\} = 1 - (A_o + A_1 + A_2) .$$

In modelling the process of irradiation, the first two coefficients of the generating function are required for cells of the r=2 level of damage and only one for those with r=3.

Realization of this scheme in the GPSS/360, in conducting simulation experiments, is accomplished in the following manner.

The program refers to the pseudo-random number generator built into the language which generate a sample of numbers distributed uniformly within the interval $[0,1)$. Depending on which of the intervals $[0,A_o)$, $[A_o,A_o+A_1)$, $[A_o+A_1, A_o+A_1+A_2)$, $[A_o+A_1+A_2,1)$ contains this number, the cell either remains at the same level of damage or moves to a deeper one.

Further redistribution of cells of an irradiated population among levels of damage is guided only by the processes of radiation damage repair.

The number of damage levels determines the dose interval the model operates in. As the irradiation dose increases, more and more cells are accumulated at the absorbing level r=4. This may lead to an abnormally high interphase cell death rate under the given conditions of irradiation.

Thus, extension of the dose range would necessitate introduction in the model of extra levels of damage which, in turn, would make the model too cumbersome. The alternative to amplifying the model with new levels of damage is the procedure of rarefying the flow of cells passing to the absorbing level. The procedure is as follows.

For each cell reaching that level a pseudo-random number distributed uniformly in the interval $[0,1)$ is generated and compared to the specified number which is one of the model's parameters. Whenever that number is above the threshold parameter, the cell moves to the level r=4, otherwise it remains at the level r=3. The distribution of cell life time at the level r=4 is exponential. Cells whose residual life span at that level has run out are eliminated from the model. The average time interval preceding the death of a level-four cell should be chosen, as a rule, equal to or shorter than the expected MC duration.

2.5. Radiation Blocking of Cells in the Mitotic Cycle

The fact of changes in the temporal parameters of the mitotic cycle after exposure to ionizing radiation has been repeatedly described in the literature. It is customary to assume that the principal contribution to the prolongation of the first post-irradiation MC is that of the delay in the transition of cells from one cycle phase to another.

Okada's monograph [52] contains an extensive review of data on radiation blocks arising in different MC periods. It follows from those data that in most of the cell lines studied in vitro there is a decline in the rate of transition of cells from the S phase to the G_2 phase and from the G_2 phase to the M phase. The most stable phenomenon is the $G_2 \rightarrow M$ block. The initiation of the G_2 block is believed to be responsible for the disturbances in the cardinal processes occurring in the G_2 phase of the MC, such as the initial processes of chromosome condensation and the formation of spindle as well as principal components of the mitotic apparatus of a cell. The duration of the G_2 block is interpreted as the time required for the elimination of at least some of such disturbances to the level sufficient for the initiation of mitosis. In publication [82] an extensive bibliography of works (28 references) is given which deal with the radiation delay of cells in the G_2 period. The authors obtained experimental data showing that the delay of HeLa cells in the G_2 phase occurs not only in case of a single exposure to acute irradiation but also under a prolonged (continuous) action of γ- rays, as well as in the case of multiple fractionation (with 12-hour intervals) of the dose with an average dose rate of 0.37 Gy/hr. The shortest duration of the G_2 block recorded by them was that of 10 hours. Under similar conditions Mitchell and Bedford [51] noted no delay of cells in the G_1 and S phases of the first post-irradiation MC. Some authors reported delay of HeLa cells in culture in the S phase of the MC under irradiation in doses of 3-5 Gy [54, 71, 82] . This phenomenon is most probably associated with a lower rate of DNA synthesis and, consequently, a longer S period, rather than with the blocking of cells on the $G_1 \rightarrow S$ boundary. Indeed, Tolmach and Jones [73] demonstrated a dose-dependent decline of the rate of DNA synthesis in the culture of HeLa S3 cells. However, for the convenience of modelling, it is expedient to introduce extra time for cell delay at the $S \rightarrow G_2$ boundary, rather than changing the temporal parameters of the S phase after irradiation. This variant was adopted in the program realization of this model as well.

In reviewing published data, Okada [52] comes to a conclusion that in rapidly dividing cells irradiation, as a rule, does not result in the $G_1 \rightarrow S$ block. The only exception noted by Okada [52] were the findings of Mak and Till [49] which extablished a decrease in the labelling index in the culture of mouse fibroblasts L-60 TM at substantial irradiation doses (5.57 Gy).

On the basis of evidence reported in the reviews of Okada [52] and Yarmonenko et al. [86] , in constructing the model, we chose a linear relationship between (non-random) durations of blocks and irradiation doses. The maximum values for block durations (with a dose of 8 Gy) were as follows: 2.3 hr for $G_1 \rightarrow S$ (in some simulation experiments the delay was assumed to be equal to 0), 4.6 hr for $S \rightarrow G_2$, and 23 hr for $G_2 \rightarrow M$. The blocks were localized on the phase boundaries over the entire range of doses, which is an obvious simplification of the actual situation. Simulation of a more general case (blocks within MC phases) presenting considerable technical difficulties was not accomplished.

The program realization of delay in the progress of cells through the MC under irradiation - the "FORMATION OF RADIATION BLOCKS" - was effected according to the following scheme. Upon completion of the operation of the "RADIATION DAMAGE" program (see the preceding section), for the cells that, as a result of irradiation, appeared to be at the r=2 or 3 levels of damage, time T_R is calculated which they would have taken to complete the MC phase they were irradiated in at the instant t_D, had there been no irradiation effect. Then the delay time T_b is computed which depends linearly on the dose (and MC phase). Cells for which $T_b > T_R$ are delayed in the phase they were damaged in for the time equal to $T_b - T_R$.

Thus, the effect of partial synchronization of cells at phase boundaries is attained; the undamaged cells (r=1 level) are continuing the MC without delay.

2.6. Repair of Radiation Damage of Cells

The formalization of the process of radiation lesion formation discussed above makes it possible to simulate the repair process by introducing reverse transitions from "deeper" levels of damage to those characterized by lesser values of probabilities of reproductive cell death. The probability of reverse transition, which is assumed to be independent of the level number (r=2,3) and whose meaning will be discussed in more detail later on, will be referred to as repair function (RF) and designated with the letter R. Of key importance in describing RF are the following points:

1. RF dependence on the time that has elapsed after irradiation.
2. RF dependence on irradiation dose.
3. RF dependence on the cell cycle stage.

It is apparent from general considerations that after irradiation RF should first increase and then, after a certain time interval, diminish, reaching gradually the initial basal level E_0 characteristic of unirradiated cells. This type of RF dependence on time E(t) is represented schematically in Figure 2.3.a. For the sake of convenience piecewise-linear approximation of RF was chosen; in this case E(t) may be determined only by specifying values E_0, t_1, t_2, t_3, t_4 and t_5. It is very important to evaluate, if only approximately, the temporal characteristics of RF on the basis of experimental data. For this purpose let us consider the results of investigations in damage repair dynamics in cell culture. There is an extensive literature devoted to the description and interpretation of observations of sublethal and potentially lethal damage repair. Sublethal damage repair (SLDR) is usually studied in fractionated irradiation experiments [15]. The most common experimental procedure for investigating potentially lethal damage repair (PLDR) is as follows: in changing the time interval between irradiation and the first stage of the method for

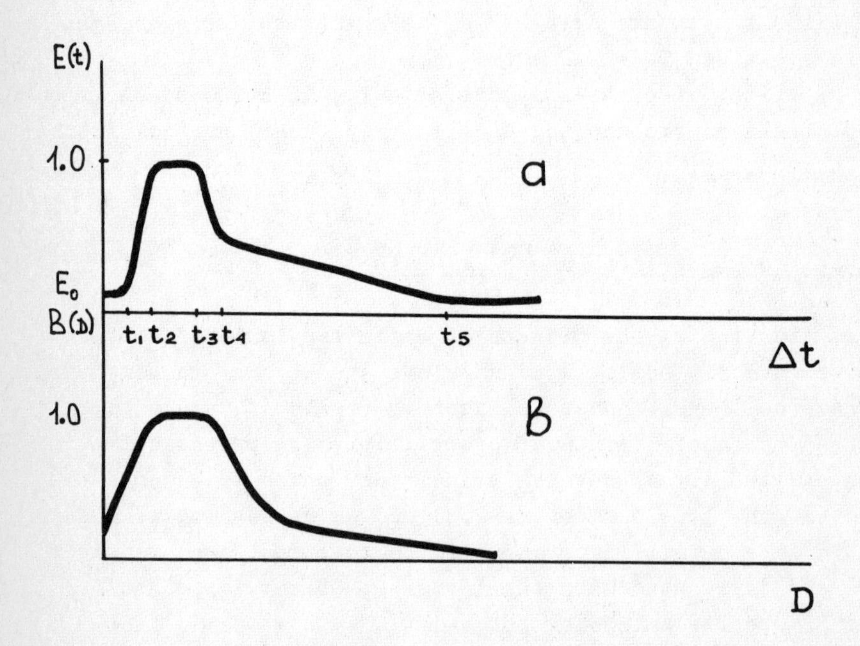

Figure 2.3 (a) Time-dependent component E(t) of repair function. $E_0, t_1, t_2, t_3, t_4, t_5$ - control points for assigning E(t).
(b) Dose-dependent component B(D) of repair function.

cell survival evaluation, cell culture trypsinization, variation (increase, as a rule) in cell survival is plotted against time that has elapsed after irradiation [13, 18, 28, 29, 46, 47, 48, 61, 78-81, 87]. A similar layout is employed in studying PLDR in ascitic and solid tumours, in the latter case the time interval being altered is that between irradiation and the preparation of tumour cell suspension [30, 36, 48, 65, 79]. It has been shown that the effect of PLDR is prominent in the cells of a culture which is in the stationary phase of growth. In exponentially growing cell cultures this effect is quite negligible [48, 61] . Little et al. [48] have found that in the plateau phase of cell culture growth, PLDR may take place after exposure to doses of 2Gy. The phenomenon of PLDR has been also revealed by determining the clonogenic capacity of tumour cells in vivo after their irradiation under the conditions of choronic and acute hypoxia [74] .

In the proposed model the processes of both SLDR and PLDR are described by a single mechanism which admits of the transition of a cell to levels of less pronounced radiation damage in all of its life cycle phases. Some investigators [84] hold the view that both types of repair are merely different manifestations of one and the same phenomenon or that, at least, they involve the same cell structures. On the other hand, Phillips and Tolmach [57] suggested that PLDR and SLDR processes are strictly dissimilar. Their opinion was afterwards shared by Utsumi and Elkind [14, 75, 76] who, with a view to identifying the PLDR component in Chinese hamster cells exposed to X-irradiation, carried out post-irradiation incubation of the cells in anisotonic saline solutions (anisotonic phosphate-saline buffer). Producing no toxic effect on the irradiated cells, such incubation led to a lowering survival rate of the cells at almost all points of the mitotic cycle, except the G_1/S boundary. The decline in the cell survival observed by the authors was dependent on the temperature and time of incubation: if incubation was started with a delay of one hour after exposure to radiation, the effect of survival-rate decline (a steeper slope of the effect-dose curve) was not apparent. Under similar experimental conditions the capacity for SLDR underwent practically no change. The authors came to the conclusion that incubation of cells in anisotonic saline solutions results in inhibition of only rapid processes of PLDR, and that PLDR and SLDR processes are independent of one another. The temporal characteristics of the process of PLD repair, obtained in studying the relationship between cell survival level and the time of the cells remain in stationary culture after irradiation, differ from the estimate (1 hr) given in the above-mentioned papers by Utsumi and Elkind. It has been shown in a large body of experimental studies [13, 28, 29, 31, 35, 36, 46, 47, 48, 56, 74, 77, 78, 79]that PLDR is generally completed in 4-6 hours. The process of SLDR has the same temporal characteristics [16, 46, 47, 78]. Only a few authors observed under similar conditions a somewhat more rapid (1-3 hr) PLDR [61]. Ordinarily, rapid processes of PLDR (with time of completion up to 1 hr) are revealed in experiments using various kinds of additional effects on cells, such as incubation at low (20°C) temperatures [84] or in a balanced saline solution [4]. Such experiments do not rule out the possibility that effects of that kind favour the development of radiation damage proper, and that PLD repair inhibition is not a characteristic feature of the effect of lowering cell survival. It is conceivable that the action of anisotonic saline solution in the experiments of Utsumi and Elkind leads, primarily, to disturbances of intracellular homeostasis, which enhances the intensity of the first component in the chain of events occurring in an irradiated cell. In this connection it is relevant to note that a lowering survival of cells incubated in anisotonic solution also prevails in the case of irradiation with fission neutrons, whereas cells exposed to ultraviolet radiation do not show this effect under similar conditions [75, 76]. In the study by Raaphorst and Dewey [60] it is demonstrated that incubation during or immediately after irradiation of CHO cells in hypertonic (1.5M) or hypotonic (0.05M) NaCl solution inhibits repair (the authors' terminology) both of SLD and PLD. Such incubation of cells prior to irradiation exerts no influence on either type of radiation damage repair.

Thus, a great number of experimental facts attest that the processes of radiation damage repair in cell cultures (and transplanted tumours) develop within time intervals of 4 to 6 hours after irradiation. This circumstance was taken advantage of when we defined the form of function E(t) represented shematically in Figure 2.3a.

The repair function-irradiation dose relationship was introduced by defining coefficient B(D) by which all curve E(t) ordinates are multiplied. Simulation experiments aimed at reproducing typical dose-effect curves (possessing a shoulder and an exponential portion) have given the principal form of the B(D) relationship given in Figure 2.3b. The B(D) relationship obtained has an apparent similitude to experimental data given in Haynes' paper [34] on changes in the number of excised dimers after cells of a wild strain of haploid yeast were exposed to different doses of ultraviolet irradiation. Simulation experiments have shown that with the chosen formalization of the processes of radiation damage and post-irradiation repair the B(D) relationship proves to be the leading factor responsible for the general form of the effect-dose curve.

The relationship between RF and cell cycle phase C(W) will be briefly discussed in the next section.

In the program realization of the process of repairing lesions which occur in irradiated cells, account was taken of the above-listed three factors determining the specific form of repair function, i.e. the efficiency of the process. The process is assumed to be of stochastic nature, and the repair function has the sense of probability that, with access to the subprogram "REPAIR",the cell would move from the r level to that of r-1, i.e. one act of repair would take place. Consequently, RF is determined in the following manner

$$R = E(t)\ B(D)\ C(W)\ .$$

After a cell completes a given phase (or sub-phase) of its life cycle or before simulating irradiation, it is examined to determine its level of damage r.

If r=1, i.e. the cell is intact, it continues its life cycle. In case the cell is at the r=2 or r=3 level, the process of damage repair is modelled for it for the period it was in the state under consideration from the instant t_1 of entering the phase till the instant t_2 of completing it. That process is modelled as follows. At a specified time interval Δt (in the model realized $\Delta t=1$ hr) a random number is generated, uniformly distributed in the interval $[0.1)$ and compared to the repair function value R calculated for the corresponding instant of model time. If the random number is below R, the cell moves from the level r to the level r-1 and if the level becomes r=1 the cell continues its life cycle. In the event of r=2 the modelling of the repair process goes on. The instants at which a cell may pass to a higher level are $t_i = t_1 + i\,\Delta t$, $i = 1,2...k$, where k is the integer part of the relation $(t_2-t_1)/\Delta t$. The process of repair is so modelled that through one act of repair the cell may move only to the neighbouring level of damage, i.e. from the level r=3 to the level r=2, not to r=1. For the complete repair of a cell of the r=3 level, two realized acts of repair are indispensable.

2.7. Relation between Cell Radiosensitivity and Its Position in Cell Cycle

Experiments with irradiation of synchronized cell populations indicate conclusively that the survival of cells may vary greatly, depending on their location in the cycle at the time of irradiation [19, 52, 56, 66, 71, 86]. The ability of cells for SLDR depends likewise upon their position in the mitotic cycle [62] . Although certain common qualitative regularities in the variations of radiosensitivity in progress through the MC have been established for many cell lines [55] every type of cells is characterized by its own specific pattern of such variations. Thus, it appears impracticable to establish once and for all any definite form of radioresistance (probability P) and repair efficiency (coefficient C(W)) dependence on the biological or chronological age W of the cell in the cycle. This relationship is prescribed for each particular type of cells based on experiments with synchronized cell culture. The question arises of whether, prescribing variations in parameters P and C(W) during the MC for a certain type of cells, one could use the model for studying cell survival in a population of another type, fitting only average (among age-groups) values of radiosensitivity and repair efficiency parameters. The simulation experiments conducted by us have provided the affirmative answer to that question, indicating that in order to obtain many radiobiological characteristics (effect-dose curve and PLDR or SLDR for a total asynchronous population) it is sufficient to prescribe cycle averaged values of P and C. However, changes in those parameters and their effect on survival-rate variations through the MC are of no little interest in themselves. For this reason the MC was divided into 19 sub-phases, with parameters P and C assumed to be constant throughout each sub-phase. As a result the G_1 period was broken into 8, the S period into 7, and the G_2 period into 3 sub-phases.

The M and G_0 periods were considered as homogenous states as regards radiosensitivity and intensity of repair processes. For each MC sub-phase the processes of development of radiation lesions and of their subsequent repair were simulated using the "RADIATION DAMAGE" and "REPAIR" programs.

2.8. Criterion of Radiation Effect

With a view to maximum approximation to the conditions of an actual radiobiological experiment for estimating the efficiency of radiation effect, use was made of the clonogenic capacity of cells defined as the ratio of the number of cells capable of cloning to the total number of cells present at the time of the onset of cloning (cell culture explant). Another parameter was also used calculated as the ratio between the number of clonogenic cells and the total number of potentially viable (localized at the r=1,2,3 levels) elements of a cell population.

Simulation of the cloning process (the program "CELL CLONES") was accomplished in the following manner. At the time of explant which is to be prescribed in advance in describing the modelling conditions, all viable G_0 cells are placed at the initial point of the G_1 phase and allowed to traverse in succession the division cycles, bypassing the G_0 phase, up to the time of "halting". This course is likewise open to cells which at the time of the onset of cloning are in the MC, they complete the cycle and their progeny continue proliferating up to the instant of "halting".

The instant of "halting" is defined as the time of taking decision on the clonogenic potential of the cell under study. Clearly, the instants of "halting" vary for individual cells tested for clonogenic capacity. A cell is assumed to be clonogenic provided : (1) it is at the r=1 level of damage at the start of testing its clonogenic capacity; (2) in the course of clone growth the cell itself (in the first MC) or one of its descendants appears at the r=1 level; and (3) as a result of this cell's reproduction a clone has formed (within a time interval not limited in advance) consisting of a certain number (50) of cells. The latter condition obviates the need for taking into account the "final-growth effect" in analyzing the damaging action of ionizing radiation, No special problems are involved in so modifying the "CELL CLONES" program that one could determine the clonogenic capacity at a fixed instant of "halting" (the same for all the cells) by counting the clones whose size exceeds a certain specified value. The model allows for "explant" of cells to test their clonogenic capicity subsequent to irradiation or after any prescribed time interval.

2.9. Fractionated Irradiation

For the purpose of simulating fractionated irradiation regimes, the proposed model was so modified that, with the time interval Δt_D between successive fractions of the dose of ionizing radiation tending to zero, the model of fractionated irradiation would degenerate into the corresponding analogue for the single irradiation with the total dose. In the event of the interval Δt_D exceeding zero, the E(t) function, which determines changes in the character of repair processes in time, should be modified too. Besides, the argument of the function B(D) (repair-dose relationship) should also be formed with due regard for the time interval between fractions.

It has not been possible to find, on the basis of published evidence, the way in which the form of the E(T) function changes after reirradiation. Under the circumstances it was reasonable to assume that, basically, the behaviour of $E(t)$ does not alter after each irradiation, although the numerical characteristics of the function do depend on the time intervals between fractions. Let t_{D_1} and t_{D_2} be instants of the first and second irradiations respectively, and Δt_D the interval of time between two successive irradiations.

The formation of the function $E(t)$ in the case of fractionated (for two fractions of a dose) irradiation in the proposed model is illustrated in Figure 2.4 where the solid line depicts the function $E(t)$ constructed for the irradiation regime with $t_{D_1} = 0$ and $\Delta t_D > 0$.

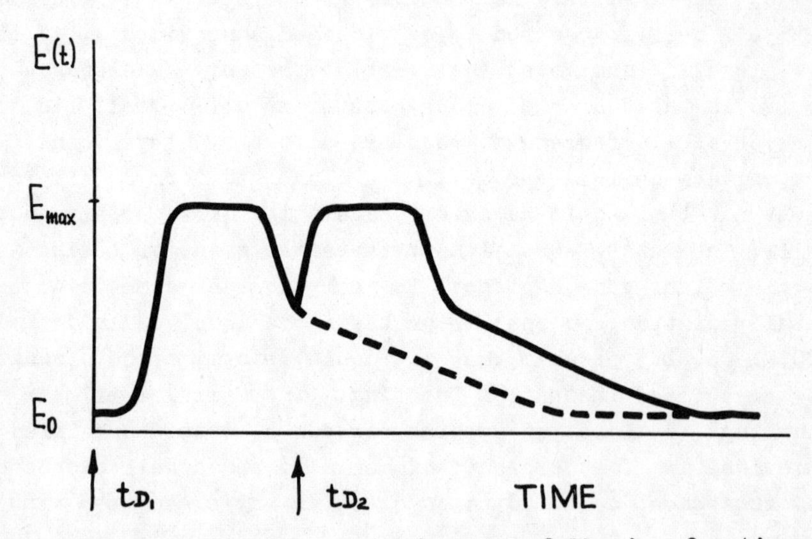

Figure 2.4. Principal form of repair function following fractionated irradiation at the instants t_{D_1} and t_{D_2}.

The dotted-line curve represents the form of $E(t)$ with a single irradiation at the instant $t = 0$.

With a view to forming the argument of the function $B(D)$ the concept of a "resulting dose" was introduced which is determined as follows:

$$D_k' = D_{k-1}' \cdot f(\Delta t_k) + D_k ,$$

where D_k is the quantity of the dose of the kth fraction, Δt_k is the time interval between (k-1)th and kth fractions of the dose, and $f(\Delta t_k)$ is the function linearly decreasing from unity to zero. The value l of the argument of the function $f(t)$, at which $f(l)=0$, will be denoted the "expiration time" for the previous dose, expressing with this term the fact that after the l period the ability to activate repair system characteristic for a given irradiation dose is fully restored.

Cells, which as a result of exposure to the 1,2...,(k-1) fractions, were in a radiation block at the time of irradiation t_{D_k} were regarded as belonging to the corresponding phase of the mitotic cycle prolonged by the value of the block duration. In the realization of the RADIATION DAMAGE sub-program the values of radioresistance P(W) for such cells were chosen according to their position in the cycle phase. In simulating the process of damage repair the values C(W) were selected in a similar manner.

It has been shown in some works [45, 72] that cells blocked in the G_2 phase retain the capacity for a further delay in division under the effect of subsequent irradiation up to their release from the radiation block created by the action of the initial irradiation. In fractionating the dose, the delay of cells in radiation blocks was so devised that with a repeated damage of cells in the block T_{k-1} (T_{k-1} is the duration of block after the fraction (k-1)), that block was prolonged, if T_k was greater than the time left before the cell's exit from the block ΔT_{k-1}, by the quantity $T_k - \Delta T_{k-1}$. For the cells whose progress through the mitotic cycle was not blocked at the time of re-irradiation, the duration of the delay was defined as in the case of a single exposure.

2.10. Information Available to Investigator in Conducting Simulation Experiments

In the course of computer-conducted simulation experiments output information on the cell system is routinely printed at those instants of model time when the process (or processes) of radiation effect on the cell system is (are) simulated prior to and right after irradiation, at the instant culture explant to test clonogenic capacity is simulated, and upon completion of modelling. Besides,information on the state of the cell system may be displayed at any prescribed intervals of time. Current information (before completion of simulation) comprises the following data on the cell system.

1. Distribution of cells over life-cycle phases and levels of damage. Total number of cells in the MC and in the whole system under simulation.

2. Number of irreversibly damaged cells (level r=4 cells) and number of such cells removed from the population by the given instant of modelling time.

3. Number of cells eliminated from the resting state in simulating the steady state (plateau-phase of growth) of cell culture.

4. Number of cells which have passed to the fourth level of damage after completing the M phase of the mitotic cycle (reproductive death).

5. Total number of repair transitions of cells (transitions from lower to higher levels of damage) and their distribution over life--cycle phases and mitotic cycle sub-phases.

At the time irradiation is being simulated, distribution of cells over the MC sub-phases is displayed in addition to information on distribution over the conventional four phases of the MC.

On completion of simulation, besides information on repair processes and the number of cells which perished reproductively or by way of the interphase type of death, the following summarized data may be printed:

1. Absolute number of clonogenic cells and their proportion both to the viable cells (levels 1,2 and 3) of the population and to the total number of viable and perishing (level 4) cells.

2. Absolute number of clonogenic cells which, at the time of "explant", were in the mitotic cycle and in the resting state, respectively.

3. Relative and cumulative frequencies for the time of the cell's residence in the resting phase as well as the sampled mean and variance of this time.

REFERENCES

1. Alper, T. Elkind recovery and "sub-lethal damage": a misleading association?, Brit. J. Radiol., 50, 459-467, 1977.
2. Alper, T. Keynote address: survival curve models, In:Radiation Biology in Cancer Research, Raven Press, New York, 3-18, 1980.
3. Barendsen, G.W. Damage to the reproductive capacity of human cells in tissue culture by ionizing radiations of different linear energy transfer, In:The Initial Effects of Ionizing Radiations on Cells, Academic Press, Lond., 183-199, 1961,
4. Belli, J.A. and Shelton, M. Potentially lethal radiation damage: repair by mammalian cells in culture, Science, 165, 490-492, 1969.
5. Bender, M.A. and Gooch, P.C. The kinetics of X-ray survival of mammalian cells in vitro, Int. J. Radiat. Biol., 5, 133-145, 1962.
6. Brown, J.M. Long G_1 or G_0 state: a method of resolvings the dilemma for the cell cycle and in vivo population. Exper. Cell Res., 52, 565-570, 1968.
7. Burns, F.J. and Tannock, I.F. On the existence of a G_0-phase in the cell cycle, Cell Tissue Kinet., 3, 321-324, 1970.

8. Chadwick, K.H. and Leenhouts, H.P. A molecular theory of cell survival, Phys. Med. Biol., 18, 78-87, 1973.
9. Cox, R. and Masson, W.K. Changes in radiosensitivity during the in vitro growth of diploid human fibroblasts, Int. J. Radiat, Biol., 26, 193-196, 1974.
10. De Maertelaer, V. and Galand, P. Some properties of a "G_0" model of the cell cycle. I. Investigations on the posible existence of natural constraints on the theoretical model in steady state conditions, Cell Tissue, Kinet., 8, 11-22, 1975.
11. De Maertelaer, V. and Galand, P. Some properties of a "G_0" model of the cell cycle. II. Natural constraints of the theoretical modeli in exponential growth conditions, Cell Tissue. Kinet., 10, 35-42, 1977.
12. Djordjevic, B., Evans, R.G., Perez, A.G. and Weill, M.K. Spontaneous unscheduled DNA synthesis in G_1 HeLa cells, Nature, 224, 803-804, 1969.
13. Dritschilo, A., Piro, A.J. and Belli, J.A. Repair of radiation damage in plateau-phase mammalian cells: relationship between sub-lethal and potentially lethal damage states, Int. J. Radiat. Biol., 30, 565-569, 1976.
14. Elkind, M.M. Cells targets and molecules in radiation biology. In: Radiation Biology in Cancer Research, Raven Press, New York, 71-93, 1980.
15. Elking, M.M. and Sutton, H.X-ray damage and recovery in mammalian cells in culture, Nature, 184, 1293-1295, 1959.
16. Elkind, M.M., Sutton-Gilbert, H., Moses, W.B. and Kamper, C. Sub-lethal and lethal radiation damage, Nature, 214, 1088-1092, 1967.
17. Elkind, M.M. and Whitmore, G.F. The radiobiology of cultured mammalian cells, Gordon and Breach, New York, Lond., Paris, 1967.
18. Evans, R.G., Bagshaw, M.A., Gordon, L.F., Kurkjian, S.D. and Hahn, G.M. Modification of recovery fromp potentially lethal X-ray damage in plateau phase Chinese humster cells, Radiat. Res., 59, 597-605, 1974.
19. Fidorra , J. and Linden, W.A. Radiosensitivity and recovery of mouse L cells, Radiat. and Environ. Bipphys., 14, 285-294,1977.
20. Garrett, W.R. and Payne, M.G. Applications of models for cell survival: the fixation time picture, Radiat. Res., 73, 201-211, 1978.
21. Ginsberg, D.M. and Jagger, J. Evidence that initial ultraviolet lethal damage in Escherichia coli strain 15 T¯A¯U¯is independent of growth phase, J. Gen. Microbiol., 40, 171-184, 1965
22. Coodhead, D.T. Models of radiation inactivation and mutagenesis, In:Radiation Biology in Cancer Research, Raven Press, New York, 231-247, 1980.
23. Goodhead, D.T., Thacker, J. and Cox, R. The conflict between the biological effects of ultrasoft X-rays and microdosimetric measurements and application, In:Proceedings of the Sixth Symposium on Microdosimetry, Commission of the European Communities, 829-843, 1978.
24. Green, A.E.S. and Burki, J.A note on survival curves with shoulders, Radiat. Res., 60, 536-540, 1974.
25. Grove, G.L. and Cristofalo, V.J. The transition probability model and the regulation of proliferation of human diploid cell cultures during aging, Cell Tissue Kinet., 9, 395-399, 1976.
26. Grove, G.L. and Cristofalo, V.J. Transition probability model and aging human diploid cell cultures, Cell, Biol. Int. Rep., 2, 185-188, 1978.
27. Gushchin, V.A. Branching of the G_1-phase in the mitotic cycle of guinea-pig colon crypt cells, Cytology, 18, 1455-1463, 1976, (In Russian).

28. Hahn, G.M. and Little, J.B. Plateau-phase cultures of mammalian cells: on in vitro model for human cancer, Curr. Topic Radiat. Res., 8, 39-43, 1972.
29. Hahn, G.M., Bagshaw, M.A., Evans, R.G and Gordon, L.F. Repair of potentially lethal lesions in X-irradiated, density-inhibited Chinese hamster cells: metabolic effects and hypoxia, Radiat. Res., 55, 280-290, 1973.
30. Hahm, G.M., Rockwell, S., Kallman, R.F., Gordon, L.F. and Frindel, E. Repair of potentially lethal damage in vivo in solid tumor cells after irradiation, Cancer Res., 34, 351-354, 1974.
31. Harris, J.-R., Murthy, A.K. and Belli, J.A. Repair following combined X-ray and heat at 41° in plateau-phase mammalian cells, Cancer Res., 37, 3374-3378, 1977.
32. Hartmann, N.P., Gilbert, C.M., Jansson, B., Macdonald, P.D.M., Steel, G.G. and Valleron, A.J. A comparison of computer methods for the analysis of fraction labelled mitoses curves, Cell Tissue Kinet., 8, 119-124, 1975.
33. Haynes, R.H. The interpretaion of mecrobial inactivation and recovery phenomena, Radiat. Res., suppl., 6, 1-29, 1966.
34. Haynes, R.H. The effect of reparation process on survival curves, In:Cell Survival after Low Doses of Radiation:Theoretical and Clinical Implications, Proceedings of the Sixth L.H. Gray Conference (T.Apler, Editor), the Institute of Physics, John Wiley and Sons, 1975.
35. Hetzel, F.W., Kruuv, J. and Frey, H.E. Repair of potentially lethal damage in X-irradiated V79 cells, Radiat, Res., 68, 308--319, 1976.
36. Hill, R.P., Warren, B.F. and Bush, R.S. The effect of intercellular contact on the radiation sensitivity of KHT sarcoma cells, Radiat. Res., 77, 182-192, 1979.
37. Hug, O. and Kellerer, A.M. Stochastik der Strahlenwirkung, Springer-Verlag, Berlin, Heidelberg, New York, 1966.
38. Kapul'tcevich, Yu.G. Quantitative regularities of cell radiation injury, Atomizdat, Moscow, 1964, (In Russian).
39. Kapul'tcevich, Yu.G. and Korogodin, V.I. Statistical models of postradiation recovery of cells, Radiobiology, 4, 349-356, 1964, (In Russian).
40. Katz, R., Ackeron, B., Hamayconfar, M. and Sharma, S.G. Inactivation of cells by heavy ion bombardment, Radiat. Res., 47, 402-425,1971.
41. Kellerer, A.M. and Rossi, H.H. The theory of dual radiation action, Curr. Topic Radiat. Res., 8, 85-158, 1972.
42. Kellerer, A.M. and Rossi, H.H. A generalized formulation of dual radiation action, Radiat, Res., 75, 471-488, 1978.
43. Laurie, J.J., Orr, S. and Foster, C.J. Repair processes and cell survival, Brit. J. Radiol., 45, 362-368, 1972.
44. Leenhouts, H.P. and Chadwick, K.H. Stopping power and the radiobiological effect of electrons, gamma rays and ions. In:Proceedings of the Fifth Symposium on Microdosimetry, Commission of the European Communities, Luxembourg, 289-308, 1976.
45. Leeper, D.B. and Hagemann, R.F. Repair kinetics of radiation-induced mitotic delay, Biophys, J., 13, 179-185, 1973.
46. Little, J.B. Repair of sub-lethal and potentially lethal radiation damage in plateau phase cultures of human cells, Nature, 224, 804-806, 1969.
47. Little, J.B. Factors influencing the repair of potentially lethal radiation damage in growth-inhibited human cells, Radiat. Res., 56, 320-333, 1973.
48. Little, J.B., Hahn, G.M., Frindel, E. and Tubiana, M. Repair of potentially lethal radiation damage in vitro and in vivo, Radiology, 106, 689-694, 1973.

49. Mak, S. and Till, J.F. The effects of X-rays on the progress of L-cells through the cell cycle, Radiat. Res., 30, 600-618, 1963.
50. Mets, T. and Verdonk, G. The theory of transition probability and division pattern of WI-38 cells, Cell Biol. Int. Rep., 2, 561-564, 1978.
51. Mitchell, J.B. and Bedford, J.S. Dose-rate effects in synchronous mammalian cells in culture. I. A comparison of the life cycle of HeLa cells during continuous irradiation or multiple-dose fractionation, Radiat, Res., 71, 547-560, 1977.
52. Okada, S., Radiation biochemistry, Volume I: Cells, Academic Press New York, Lond., 1970.
53. Orr, J.S., Wakerley, S.E. and Stark, J.M. A metabolic theory of cell survival curves, Phys. Med. Biol, 1, 103-108, 1966.
54. Painter, R.B. and Robertson, J.S. Effect of irradiation and theory of role of mitotic delay on the time course of labelling of HeLa S3 cells with tritiated thymidine, Radiat. Res., 11, 206-217, 1959.
55. Payne, M.G. and Garrett, W.R. Some relations between cell survival models having different inactivation mechanism, Radiat. Res., 62, 382-394, 1975.
56. Pelevina, I.I., Afanas'ev, G.G. and Gotlig, V.Y. Cell factors in tumour reaction to irradiation and chemotherapy, Nauka, Moscow, 1978 (In Russian).
57. Phillips, R.A. and Tolmach, L.J. Repair of potentially lethal damage in X-irradiated HeLa cells, Radiat. Res., 29, 413-432, 1966.
58. Powers, E.L. Considerations of survival curves and target theory, Phys. Med. Biol., 7, 3-28, 1962.
59. Puck, T.T. and Marcus, P.I. Action of X-rays on mammalian cells, J. Exper. Med., 103, 653-666, 1956.
60. Raaphort, G.P. and Dewey, W.C. A study of the repair of potentially lethal and sublethal radiation damage in Chinese hamster cells exposed to extremely hypo-or-hypertonic NaCl solutions, Radiat, Res., 77, 325-340, 1979.
61. Raju, M.R., Frank, J.P., Bain, E., Trujillo, T.T. and Tobey R.A. Repair of potentially lethal damage in Chinese hamster cells after X and -irradiation, Radiat. Res., 71, 614-621, 1977.
62. Schaer, J.C. and Ramseier, L. Studies on the division cycle of mammalian cells. X-ray sensitivity and repair capacity of synchronous by dividing murine mastocytoma cells, Radiat. Res., 56, 259-270, 1973.
63. Shields, R. Transition probability and the origin of variation in the cell cycle, Nature, 267, 704-707, 1977.
64. Shields, R. and Smith, J.A. Cells regulate their proliferation through alterations in transition probability, J. Cell Phyiol., 91, 345-356, 1977.
65. Shipley, W.U., Stanley, J.A., Contenay, V.D. and Fields, S.B. Repair of radiation damage in Leuis carcinoma cells following in situ treatment with fast neutrons and - rays, Cancer Res., 35, 932-938, 1975.
66. Sinclair, W.K. Sensitization by hydroxyurea and protection by cysteamin of Chinese hamster cells during the cell cycle, In: Radiation Protection and Sensitization, Tailor and Francis LTD, Lond., 201-210, 1970.
67. Sinclair, W.K. N-ethylmaleimide and the cyclic response to X-rays of synchronous Chinese hamster cells, Radiat. Res., 55, 41-57, 1973.
68. Smith, J.A. Application of the theory of transition probability in "ageing" WI-38 cells: similar behaviour of clonogenic cells from early and late passage cultures, Cell. Biol. Int. Rep., 1,283-289,1977.

69. Smith, J.A. and Martin, L. Do cells cycle?, Proc. Nat. Acad. Scien. USA, 70, 1263-1267, 1973.
70. Svetina, S. and Zeks, B. Transition probability models of the cell cycle exhibiting the age distribution for cells in the intermediate state of the cell cycle, In:Biomathematics and Cell Kinetics, Biomedical Press, Amsterdam New York, Oxford, 71-82, 1978.
71. Terasima, T. and Tolmach, L.J. Variation in several responses of HeLa cells to X-radiation during the division cycle, Biophys, J., 3, 11-33, 1963.
72. Tolmach, L.J., Griffiths, T.D. and Jones, R.W. Susceptibility of X-ray-arrested HeLa S3 cells to additional arrest, Radiat. Res., 66, 649-654, 1976.
73. Tolmach, L.J. and Jones, R.W. Dependence of the rate of DNA synthesis in irradiated HeLa S3 cells on dose and after exposure, Radiat. Res., 69, 117-133, 1977.
74. Urano, M., Nesumi, N., Ando, K., Koike, S., Ohnuma, N., Repair of potentially lethal radiation damage in acute and chronically hypoxic tumor cells in vivo, Radiology, 118, 447-451, 1976.
75. Utsumi, H. and Elkind, M.M. Potentially lethal damage. Qualitative differences between ionizing and non-ionizing radiation and implications for "single-hit" killing, Int. J. Radiat. Biol., 35, 373-380, 1979.
76. Utsumi, H. and Elkind, M.M. Potentially lethal damage versus sublethal damage: independent repair processes in actively growing Chinese hamster cells, Radiat. Res., 77, 346-360, 1779.
77. Weichselbaum, R.R. and Little, J.B. The differential response of human tumours to fractionated radiation may be due to a post--irradiation repair process, Brit. J. Cancer, 46, 532-537, 1982.
78. Weichselbaum, R.R. and Little, J.B. Repair of potentially lethal X-ray damage and possible applications to clinical radiotherapy, Int. J. Radiat. Oncol. Biol. Phys., 9, 91-96, 1983.
79. Weichselbaum, R.R., Nove, J. and Little, J.B. Radiation response of human tumor cells in vitro, In: Radiation Biology in Cancer Research, Raven Press, New York, 345-351, 1979.
80. Weichselbaum, R.R., Malcolm, A.W. and Little, J.B. Fraction size and the repair of potentially lethal radiation damage in human melanoma cell line, Radiology, 142, 225-227, 1982.
81. Weichselbaum, R.R., Schmit, A. and Little, J.B. Cellular repair factors influencing radiocurability of human malignant tumors, Brit. J. Cancer, 45, 10-16, 1982.
82. Weiss, B.G. Perturbations in precursor incorporation into DNA of X-irradiated HeLa S3 cells, Radiat. Res., 48, 128-145, 1971.
83. Wideröe, R., A comparison of radiation effects on mammalian cells in vitro caused by X-rays, high energy neutrons and negative pions, Radiat. Environ. Biphys., 15, 57-75, 1978.
84. Winans, L.F., Dewey, W.C. and Dettor, C.M. Repair of sublethal and potentially lethal X-ray damage in synchronous Chinese hamser cells, Radiat. Res., 52, 333-351, 1972.
85. Yakovlev, A.Yu., Zorin, A.V. and Isanin, N.A. The kinetic analysis of induced cell proliferation, J. Theor. Biol., 64, 1-25, 1977.
86. Yarmonenko, S.P., Wainson, A.A., Kalendo, G.S. and Rampan, Yu.I. Biological bases of radiation therapy of tumours, Medicine, Moscow, 1976, (In Russian).
87. Zinninger, G.F. and Little, J.B. Fractionated radiation response of human cells in stationary and exponentialp phases of growth, Radiology, 108, 423-428, 1973.
88. Zorin, A.V., Gushchin, V.A., Stefanenko, F.A., Cherepanova, O.N. and Yakovlev, A.Yu. Computer simulation of kinetics of irradiated cell populations in tumours, Experimental Oncology, 5,27--30, 1983, (In Russian).

III. SIMULATION AND ANALYSIS OF RADIOBIOLOGICAL EFFECTS IN CELL CULTURES

3.1. Introduction

In the preceding chapter we proposed a simulation model of the effect of ionizing radiation on in vitro cell systems. The adequacy of a multiparameter model to the actual processes for whose description it has been developed may be reliably established only when it proves instrumental in reproducing a whole set of results obtained in biological experiments.

It is along these lines that we intend to pursue the inquiry in this chapter, relying on published experimental data obtained from studying the effect of ionizing radiation on synchronized, exponentially growing and stationary cell cultures. The data belong mainly to one group of authors which is important for ensuring the uniformity of the material under consideration. At the same time these data are fairly representative, being related to research in practically all known phenomena of cell radiobiology on a specific cell line.

Computer simulation of biologic processes is an experimental activity of the researcher. For this reason a detailed description of experimental arrangement is of fundamental importance. Each section of this chapter begins with such a description which specifies parameter values and types of functional dependences introduced into the model. To avoid textual repetitions the description of each concrete experiment defines only such conditions that differ from those previously used on account of the central idea of the experiment. Wherever misunderstanding may arise references are made to the preceding sections of the monograph . Certain characteristics of the model were described in Chapter II, making it unnecessary to dwell on them at length again. Figures 3.2a and 3.2b in this Chapter represent plots of radioresistance P(W) and repair intensity C(W) changes in the course of the MC. It has been found in a special set of simulation experiments that at fixed $\bar{P}(W)$ and $\bar{C}(W)$ values obtained from averaging the P(W) and C(W) functions over the mitotic cycle the survival curves for cells irradiated in the exponential growth phase are scarcely sensitive to changes in the shape of the P(W) and C(W) dependences. Therefore the character of alterations in the radiosensitivity and repair activity of a cell as it proceeds in the MC is assumed to be identical in all the simulation experiments and interpretation of the simulation results is carried out only in terms of the averaged $\bar{P}(W)$ and $\bar{C}(W)$ values.

Within each section of this chapter, e.g. that dealing with exponentially growing or stationary cell culture, a similar situation prevails: some parameter values are taken to be equal for all the simulation experiments treated in the section. Among such parameters fixed for a large set of experiments are numerical characteristics of MC phase durations (with the exception of experiment 3 which is a special study of the role played by the average MC duration in irradiated cell survival) and radiation block durations (see Section 2.5) at MC phase boundaries at different irradiation doses. The choice of the parameter values is based on the analysis of published data on the kinetics of cell populations in the cultures concerned, the results of the analysis being summed up in [6]. Discussion of the biological purport of simulation experiment results (wherever necessary) follows immediately after their description and comparison to real observations. All the simulation experiments discussed in this chapter are numbered consecutively.

There is another aspect in the work of a specialist in simulation modelling that is no less important than a clear-cut definition of experimental conditions. That is statistical processing of the results of testing the model. Problems of that kind sometimes arise even prior to simulation trials since formulation of certain problems of modelling is intimately related to working out an optimal design of simulation experiments. What is primarily implied is the problem of revealing cellular system characteristics more significant in relation to the resultant radiation effect. An example of applying methods of factorial variance analysis to solving the problem is given in Section 3.6.

The problem of statistical analysis of random trajectories (realizations) of the simulated models presents quite a few difficulties which are of interest to experts designing software for stochastic s systems of simulation modelling. Of little use in solving that problem are the GPSS service programs directed only towards obtaining current estimates of such characteristics as the average queue length or the average time of transaction expectation. Incidentally, it is impracticable to establish by means of these programs the interrelation between modelling time and the required level of accuracy of statistical estimates. Wherever simulation of a stochastic system is aimed at estimating one of its stationary characteristics, regeneration approach proves to be an effective means of statistical analysis [4, 8] . The following are the basic requirements for applying the regeneration approach: (1) the process being simulated returns repeatedly to some fixed state (or area); (2) the average time between returns is finite; (3) the moment the process returns to the fixed state is its regeneration point. A regenerating process can be regarded as consisting of independent similarly distributed cycles (regeneration cycles), which permits the use of well developed probabilistic models in analyzing such processes.

Irradiated cell survival curves considered in this chapter represent random processes essentially non-stationary (non-homogeneous) on a certain time interval, their values being of primary interest on that particular interval. Those processes do not exhibit the property of regeneration, and due to the complexity of the model the researcher can neither construct for them a siutable probability space nor successfully simplify their structure. In our case simulation aims at reproducing radiobiologic regularities rather than estimating any stationary parameter of a cell system. Therefore simulated curve realizations have to be handled in the same way radiobiologists handle real cell survival curves. This is to say that they should be ordinate-averaged, and interval estimates (point-to-point confidence intervals) should be constructed for the sections of the process covered by experimentali information. Clearly such processing of data does not permit conclusions to be drawn concerning the stochastic relationship between different sections of the process. Wherever simulation time is not a limiting factor, a more profound analysis, e.g. estimation of the correlation function of the process under study, is practicable. However, it is very difficult to achieve with reference to the real data of a radiobiologic experiment because of the small body (sample size) of such data. The foregoing is also quite true for the estimation of mutual correlation function in comparing model realizations with the results of a radiobiologic experiment. Moreover, the correlation function of a non-stationary process depends not on one but on two arguments which cannot but complicate interpretation of statistical analysis results. The goodness of fit evaluation is of particular difficulty if one considers the problem from the viewpoint of the stochastic processes theory. This difficulty is due to the lack of succifiently general methods for constructing statistical tests to compare two non-stationary stochastic processes of arbitrary structure. The apparently fruitful idea of using for that purpose the methods for solving the multidimensional Monge-Kantorovich problem [20] was offered to us by S.T.Rachev. However, it is not yet clear how that idea could be translated into a concrete statistical criterion for comparing two random processes. Thus, here again, all we can do is point-by-point comparison of sampled data obtained from using the model with those of real experiment. To this end effective use may be made of the two-sample non-parametric criteria for testing the homogeneity hypothesis (the χ^2 test, the Kolmogorov-Smirnov test, rank tests, etc.). It should be always kept in mind however, that the resulting statistical inference is valid only for a given section of the processes being compared. In processing the simulation experiments described below, no advantage was taken of the above-mentioned tests inasmuch as the publications on which we relied for real observations contained only mean values, no information being given on original samples. Comparison of mean values may be effected with the Student-Fisher test, although in most of our cases this procedure would be quite unnecessary since a visual estimation left no doubt of good agreement between simulation results and real observations.

Offering certain prospects is the non-parametric regression approach. Indeed, considering regression dependences of simulated $Z_1(t)$ and really observed $Z_2(t)$ processes on current time, non-parametric estimators of these dependences can be constructed by pairs of observations $\{Z_1^{(i)}, t^{(i)}\}$ and $\{Z_2^{(i)}, t^{(i)}\}$, $i=1,2...,n$ and then testing the statistical hypothesis of two regression curves equality may be sought. The values $\{t^{(i)}\}$, $i=1..,n$, represent the lattice of nodes on which the values $Z_1^{(i)}$ and $Z_2^{(i)}$ are defined. For restoring regression relationships non-parametric estimators of the kernel type proposed by Nadaraya [16, 17] and Watson [26] can be used as well as the local approximation method [10]. The Nadaraya-Watson estimators are oriented to problems with random nodes distributed with a certain probability density $f(t)$. Based on those estimators, Nadaraya [17] obtained a confidence region for regression curve and constructed a statistical test for comparing two regression curves. However, those results are of asymptotic character, and the limited sample sizes, especially as regards real observations, prevent effective application of the results to comparison of simulated data with those from biologic experiments. At the same time we are convinced that the advantages of non-parametric regression analysis have not as yet been fully appreciated by simulation modelling experts and that it is bound to find extensive application in this field.

As for the accuracy of constructing model curves, its attainment by estimating the required number of model runs (simulation time) is a topical problem wherever the model is to make a prediction which will serve as a basis for some decisions. In our case, using the radiobiologic data available in the literature, we could not control their accuracy and, consequently, the problem of ensuring a given accuracy of modelling results was irrelevant. Discussed in Chapters IV and V are certain problems associated directly with studies on variance of simulated system trajectories and assessment of accuracy of cell kinetics indices.

On the basis of the foregoing we have taken the following stand regarding the methodology of employing the simulation model proposed in the preceding chapter. The principal aim of our investigation is conceptual interpretation by means of a model of data from radiobiologic experiment, the model being regarded as an independent subject of research equivalent to the biological original. To ascertain the properties of that subject of research (model) experimental approaches are applied identical to those adopted in cell radiobiology. In other words, the plan of testing the model and the form of presenting the results conform to the experimental pattern traditional for cell radiobiology. However, as a subject of investigation a model holds much wider potentialities in accounting for the peculiarities of its behaviour. It is these potentialities that we primarily intend to take advantage of for gaining further insight into radiation injury and postradiation recovery of biologic objects.

Before turning to the discussion of concrete simulation experiments let us briefly touch upon the use of such terms as "sublethal damage repair" (SLDR) and "potentially lethal damage repair" (PLDR). As pointed out in Chapter II these types of repair differ only in the ways of their detection: the recording of SLDR involves the use of fractionated irradiation whereas that of PLDR requires variation of the time interval between a single irradiation and explantation of cells to determine their clonogenic capacity. This is not an adequate classification if for no other reason than the fact that with fractionated irradiation of a stationary cell culture PLDR cannot but contribute to the resulting effect recorded by the survival of irradiated cells. Besides, there is at present no conclusive evidence that would make it possible to differentiate by mechanisms between these two types of radiation damage. We regard SLDR and PLDR as a single damage repair process whose concrete manifestations depend upon experimental design. The simulation model described in Chapter II was constructed accordingly and its subsequent application proved the fundamental validity of our concepts.

However, with a view to making the material that follow available to the widest possible readership, radiobiologists included, we have retained, in discussing the results of simulation experiments, the conventional classification. This should give no rise to misunderstanding, keeping in mind the principles of simulating post-irradiation cell repair described in the foregoing.

3.2. Irradiation of the Synchronous Cell Population

Quantitative studies of variations in the survival of cells depending on their location in the mitotic cycle at the time of irradiation are conducted in experiments with artificially synchronized cell populations. In radiobiologic experiments the most common method of synchronization is that of selecting cells in mitosis. Indeed, in the mathematical or simulation modelling of the dynamics of a population thus synchronized, it has to be assumed that all the cells at the starting time $t=0$ are at the zero age (onset) of the G_1 phase of the mitotic cycle. Subsequent desynchronization of the cell population may be studied by recording the portions of cells in the cycle phases at successive instants $t>0$. It goes without saying that a more detailed picture can be obtained with the mitotic cycle broken down into a greater number of sub-phases, as in the case of the proposed model.

Shown in Fig.3.1 are time varying distributions of cells previously synchronized for 19 sub-phases of the mitotic cycle with the following values of the temporal parameters of the G_1, S, G_2 and M phases: $\bar{\tau}_{G_1}= 8.0$, $\sigma_{G_1}= 1.5$; $\bar{\tau}_S= 7.0$, $\sigma_S= 1.3$; $\bar{\tau}_{G_2}= 3.0$, $\sigma_{G_2}= 0.9$; $\bar{\tau}_M= 1.0$, $\sigma_M= 0.5$ where τ is the mean value and σ is the standard deviation of the phase duration (in hours).

The durations of the sub-phases of the divided mitotic cycle were assumed to be independent random variables whose mean values and variances were so selected that the temporal parameters of the phases G_1, G_2, S and M would take on the values given above. As may be seen from Figure 3.1, the process of desynchronization of the cells passage through the mitotic cycle may be very speedy, and this has to be taken into account in ascertaining true variations in radioresistance P(W) and repair efficiency C(W) depending on the cell age in the cycle. Definitions of the functions P(W) and C(W) are given in Chapter II.

In principle, by choosing corresponding P(W) and C(W) relationships, one can always achieve a satisfactory reproduction of the

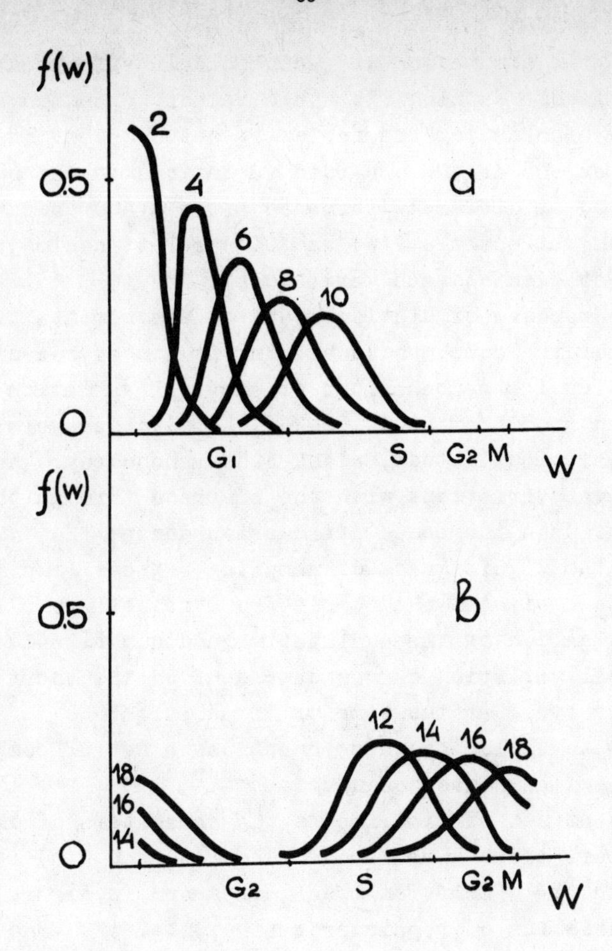

Figure 3.1. Variation in time of cell distribution through the mitotic cycle (19 subphases) in a population synchronized by selection in mitoses.
(a) Distribution 2,4,6,8 and 10 hours after synchronization.
(b) Distribution 12,14,16 and 18 hours after synchronization.

observed variations in the survival of cells irradiated at different instants following the removal of the synchronizing agent. However, it has been pointed out time and again that plotting such curves by a suitable choice of numerous model parameters is not among the principal tasks of the simulation of radiobiologic effects observed in a cell culture. In our opinion, it would be more expedient to use the model here for elucidating the following set of questions:

1. To what extent do the observed variations in the survival of cells irradiated at different instants after synchronization of the cell population reflect the P(W) and C(W) relationships responsible for the manifestation of such variations?

2. Can the proposed simulation model be instrumental (i.e. within the framework of the concepts underlying it) in reproducing qualitative regularities in the behaviour of survival variation curves for cell irradiated at different stages of the mitotic cycle? Such qualitative regularities include, among other phenomena, the greater range of survival variations with the progress through the mitotic cycle attendant upon increased irradiation dosage [24] .

3. What is the significance of sampling effects (i.e. effects due to the restricted size of the sample from the cell population) as regards investigation of the qualitative and quantitative peculiarities of survival variation curves dependent on the location of the cell within the cycle at the time of irradiation?

With a view to exploring these questions a set of computer-aided simulation experiments was conducted.

In the present set of experiments the parameters of the mitotic cycle phases were those given above. The P(W) and C(W) relationships used in testing the simulation model are shown in Figure 3.2, while other characteristics of repair processes, i.e. the functions E(t) and B(D) corresponded to those exhibited in Figures 2.3a and 2.3b of the preceding Chapter. The $\bar{P}(W)$ and $\bar{C}(W)$ values averaged over the mitotic cycle yielded the following characteristics: $\bar{P}(W)=0.81$; $\bar{C}(W)=0.14$. The durations of the blocks $G_1 \rightarrow S$, $S \rightarrow G_2$ and $G_2 \rightarrow M$ at a dose of 3 Gy were respectively : 0 hr, 1.7 hr and 8.6 hr, and at a dose of 6 Gy : 0 hr, 3.4 hr and 17.2 hr.

Cell survival variation was studied as dependent on the time interval between the synchronization of the cell population and irradiation with different radiation doses. The effects of the following radiation doses were simulated: 1Gy, 3Gy, 5Gy and 7 Gy. Irradiated cell populations differing in size were used for the above doses, viz. 10^2, 10^3, $3 \cdot 10^3$ and $6 \cdot 10^3$, respectively.

The experimental findings presented in Figure 3.3 are in agreement with the results obtained by Sinclair [24] on Chinese hamster cell culture. Investigation into the character of the survival curves shown in Figure 3.3 and their comparison to the forms of the

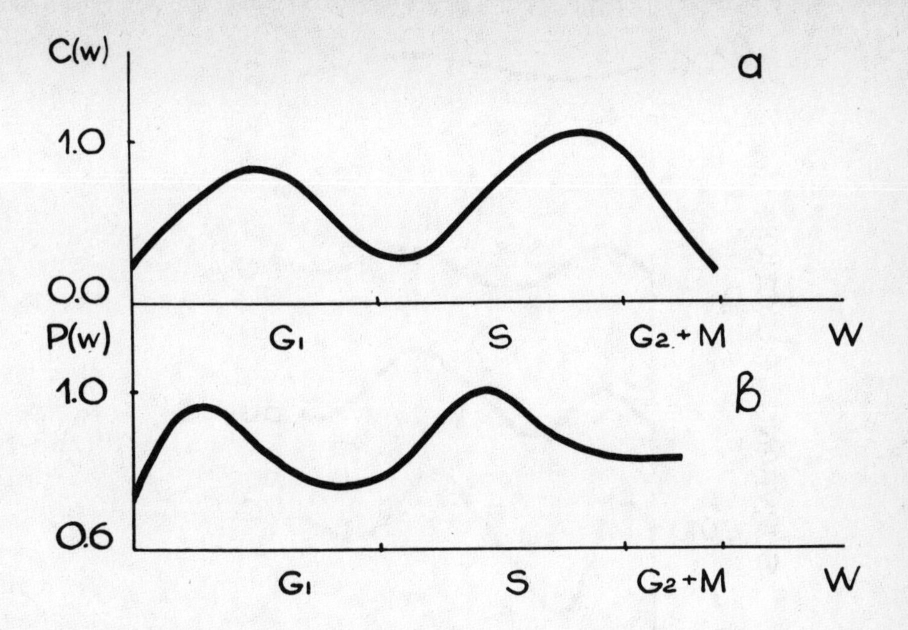

Figure 3.2. Variations depending on cell position in the mitotic cycle:
(a) of repair function,
(b) of radiosensitivity.

functions displayed in Figure 3.2 suggest the following conclusions:

1. Within a time interval shorter than the mean duration of the mitotic cycle, variations in the survival of cells irradiated at different intervals after synchronization reflect, on the whole, the character of changes in radioresistance P(W) and repair efficiency C(W) as cells procede through the cycle. Our simulation experience also shows that for not sine-like dependences of P(W) and C(W) it is impossible to reproduce sine-like survival of a synchronized population.

2. The simulation model reproduced the increasing range of variations in the survival of synchronized cells with an increase in the irradiation dose.

It should be noted that the latter conclusion can be made without involving the assumption of the existence of differences between the phases (sub-phases) of the mitotic cycle with respect to the relationship between repair efficiency and irradiation dose, i.e. the function B(D). In the simulation experiments described below this rela-

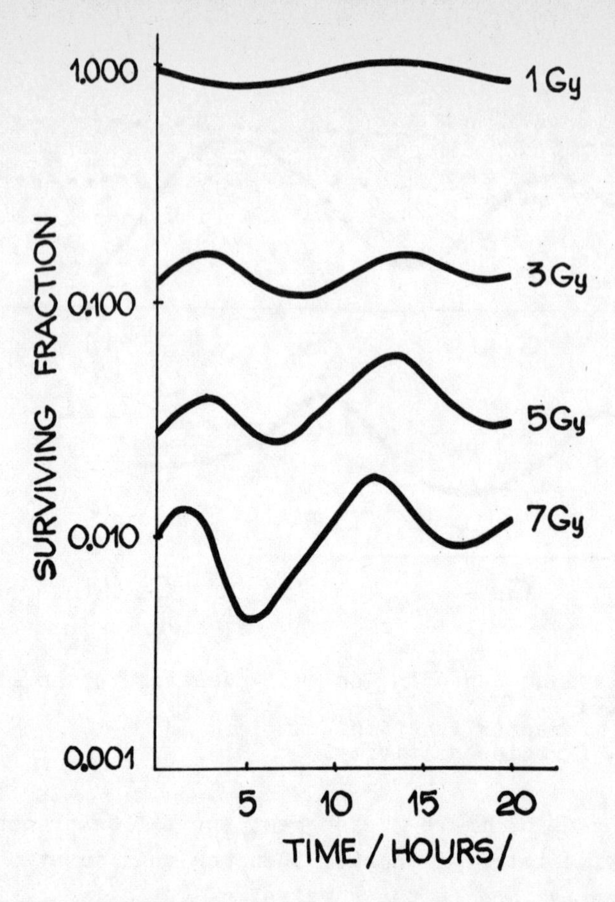

Figure 3.3. Results of simulation experiments revealing cell survival variations depending on time of irradiation elapsing after synchronization of cell population and on radiation dose.

tionship was assumed to be identical for all the phases (sub-phases) of the mitotic cycle.

Figure 3.4 demonstrates three (out of ten) realizations of survival variations for a synchronized population of 125 cells and the corresponding average over ten realizations. These data illustrate the significance of sampling effects as regards investigation of variations in the radiation damage of cells in the course of their progress through the mitotic cycle. Fluctuations recorded on small smaples (125 cells) may show a seeming increase in the range of variations. It should be noted, however, that even in the case of such a small population the location of extreme points on all of the ten realizations of the survival curve is quite stable to sampling fluc-

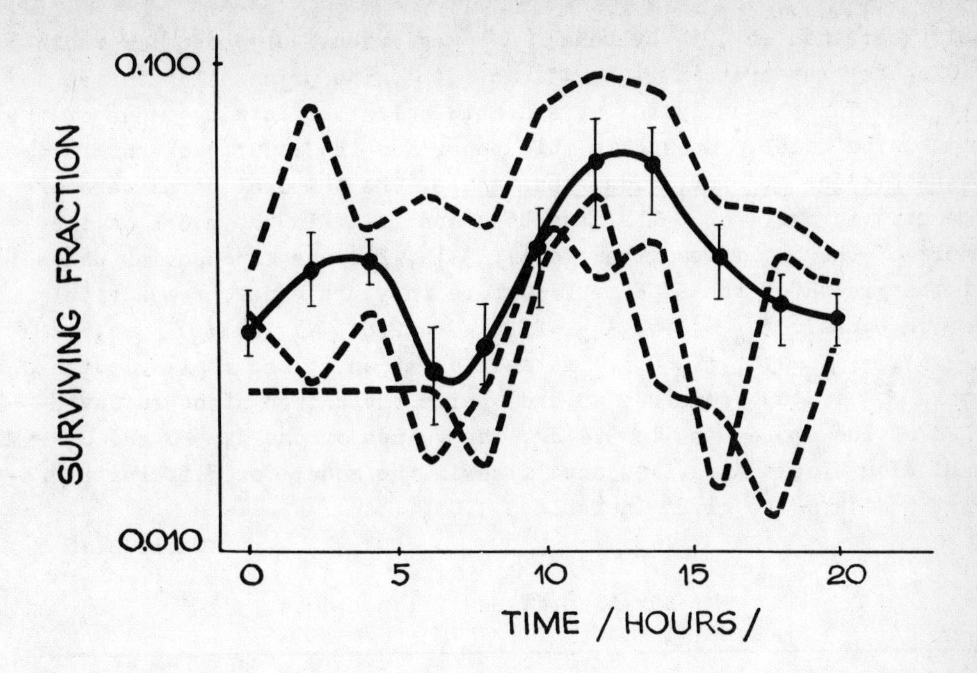

Figure 3.4. Three separate realizations (dashed lines) of survival variations for a synchronized population consisting of 125 cells. Solid line joins mean values for 10 realizations. Vertical lines - $2\sigma_{\bar{s}}$, $\sigma_{\bar{s}}$ - standard error of arithmetic mean $\bar{s}$.

tuations. Thus the type of radiobiological experiments described in this section calls for close control over the size of the cell population exposed to irradiation which, under the conditions of a radiobiological experiment, should be so chosen as to ensure the equality, at all doses, of at least average values (for the cycle and for realizations) of clonogenic cell numbers.

3.3. The Effect of Ionizing Radiation (single Irradiation) on a Cell Culture in the Exponential Phase of Growth

This section is devoted to the interpretation of data obtained from experimental studies of the effect of X-rays (220 kV) on exponentially growing cultures of human liver cells (LICH). This line of cells obtained in 1954 by Chang [3] was extensively used for radiobiological research in [12, 13, 15, 34]. The principal material utilized for the analysis of radiobiological effects by means of the simulation model proposed in this paper was the results of work [34].

Conditions of simulation experiments. The temporal parameters of the mitotic cycle of cells have been specified in the model on the basis of the estimates given in [7, 33]. For the exponential phase of the growth of the LICH cell culture they are equal, respectively, to (in hours): $\bar{\tau}_{G_1}=10.5$, $\sigma_{G1}=2.2$; $\bar{\tau}_s=12.0$, $\sigma_s=2.3$; $\bar{\tau}_{G_2}=4.5$, $\sigma_{G2}=1,2$; $\bar{\tau}_M=1.0$, $\sigma_M=0.5$. The mean duration of the whole mitotic cycle $\bar{\tau}_c$ is thus equal to 28 hrs, while the chosen standard deviation of the cycle σ_c is 3.4 hr. The values of the $G_1 \rightarrow S$ and $G_2 \rightarrow M$ radiation blocks durations specified in the model for different irradiation doses are given in Table 3.1.

Table 3.1

Duration of radiation blocks

Irradiation dose (Gy)	Duration of blocks = T (hour)	
	$T_{G1 \rightarrow S}$	$T_{G2 \rightarrow M}$
3	3.4	8.6
4	3.7	9.3
5	4.0	10.0
6	4.3	10.7
7	4.6	11.4

The values of the $S \rightarrow G_2$ blocks durations within the same dose range have been assumed to be equal to those for the $G_1 \rightarrow S$ blocks. The quantities listed in Table 3.1 were identical in all the series of experiments described in this Chapter. The function E(t) corresponded to that depicted in Figure 2.3a.

Simulation experiment 1. In this experiment the values of radiosensitivity $\bar{P}(W)$ and repair efficiency $\bar{C}(W)$ averaged over the mitotic cycle were estimated by fitting the realizations of the dose-effect simulation curves to the experimental data adopted from publication [34] and represented in Figure 3.5. The values of estimators proved to be equal to $\bar{P}(W)=0.84$, $\bar{C}(W)=0.14$. The results of simulating the dose-effect relationship obtained with the values of model parameters given above are shown in Figure 3.5. The component of the re-

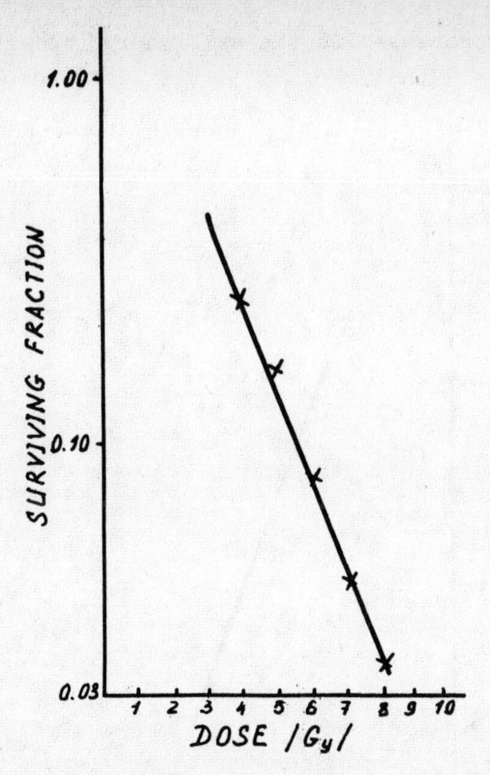

Figure 3.5. Survival of LICH cell culture irradiated in the logarithmic phase of growth. Solid line corresponds to results of simulation for $\bar{\tau}_c$ = 28 hrs. Experimental points from [34] are shown.

lationship between the repair function and the irradiation dose B(D) was also determined in this experiment (Figure 2.3b), the relationship being used in all subsequent series of computer-based experiments.

Simulation experiment 2. The purpose of this experiment was to ascertain the response of the model whenever the cloning of cells begins not right after irradiation but after a certain time interval $\Delta t_{clon.}$. It is precisely in this way that potentially lethal damage repair (PLDR) is studied in cell radiobiology . The results of the simulation experiment indicate that no change in the survival of an exponentially growing cell culture ($\bar{\tau}_c$ = 28 hr) occurs when explantation of cells is delayed for $\Delta t_{clon.}$ = 6 hours after irradiation. Illustrated in Figures 3.5 and 3.6 these results are in full agreement with the experimental findings attesting to the absence of the PLDR effect in the cells of a culture which is in the logarithmic growth phase [18, 34] .

Simulation experiment 3. Analyzing their own data obtained in studying proliferative processes in the stationary (density-inhibited) culture of LICH cells, Zinninger and Little [33], as well as Hahn and

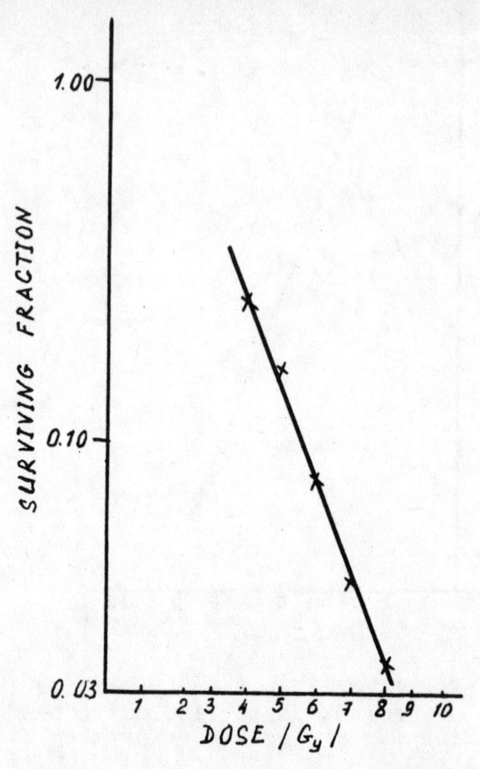

Figure 3.6. Dose-dependent survival of LICH cells for $\Delta t_{clon.}$=6 hrs and $\bar{\tau}_c$=49 hrs. Solid line-simulation results, x - experimental data.

and Little [7] , have come to a conclusion that the plateau phase of the growth of that culture is characterized by an almost twofold increase in the duration of the mitotic cycle. The authors see no reason to distinguish within that cell system a special population of cells resting outside the mitotic cycle, i.e. a population of G_0 cells. Thus, it follows from the interpretation by the said authors that the kinetics of proliferative processes in a stationary cell culture do not differ from those prevailing in the exponential state, except for some substantial differences in the duration of the cycle phases. Choosing on the grounds of the authors' recommendations the following temporal parameters of the cell cycle in the stationary phase of the culture's growth: $\bar{\tau}_{G_1}$ = 20.0, σ_{G_1}=4.0; $\bar{\tau}_S$=20.0, σ_S=4.0; $\bar{\tau}_{G_2}$=8.0, σ_{G_2}=2.0; $\bar{\tau}_M$=1.0, σ_M=0.5 , we have ascertained in the simulation experiment that with these values of the parameters the effect-dose

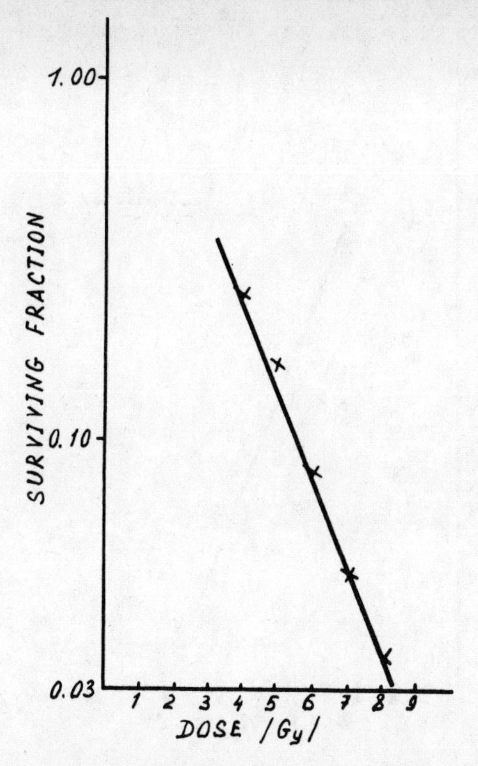

Figure 3.7. Dose-dependent survival of LICH cells for Δt_{clon}=0 hrs and $\bar{\tau}_c$=49 hrs. Solid line-simulation results, x - experimental data.

curve coincides with the similar curve for cells with ashort cycle ($\bar{\tau}_c$=28 hr) illustrated in Figure 3.7. A delay in cloning for Δt_{clon}= =6 hours for a cell culture with a prolonged mitotic cycle duration ($\bar{\tau}_c$=49 hr) neither produced any changes in the survival of those cells throughout the entire range of radiation doses under study (Figure 3.8).

The results of the three simulation experiments may be summed up as follows :

- the duration of the mitotic cycle of cells in the exponential phase of a cell culture growth exerts no influence on their survival after a single irradiation;

- a delay in the explantation of cells (to determine their clonogenic capacity) reveals no effect of PLDR, if exposed to irradiation is an exponentially growing cell population (regardless of the mitotic cycle duration).

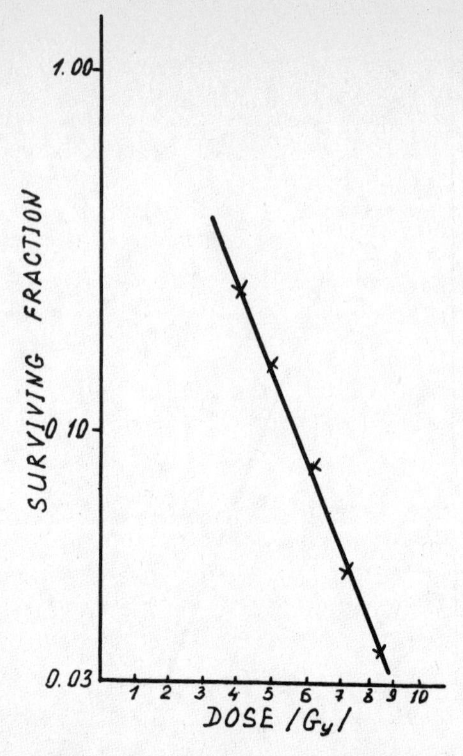

Figure 3.8. Dose-dependent survival of LICH cells for Δt_{clon}=6 hrs and $\bar{\tau}_c$=49 hrs. Solid line-simulation results. × - experimental data.

3.4. Stationary State of LICH Cell Cultures

Further analysis of the kinetics of cell populations [6] performed on the basis of published experimental data [7, 34] has shown that in the plateau phase of the growth of a LICH cell culture about 70 percent of the cells may be assumed to be in the G_0 phase. It has also disclosed that the G_2 phase of the mitotic cycle of proliferating cells has a ramified structure, evident from the temporal parameters of the G_2 phase taking on the following values with the probability 0.25 : $\bar{\tau}_{G_2}$ = 4.5 hr, σ_{G_2} = 1.2 hr, and with the probability 0.75 : $\bar{\tau}_{G_2}$ = 20.0 hr, σ_{G_2} = 5.0 hr. The temporal characteristics of the other phases of the cycle of proliferating cells were as follows: $\bar{\tau}_{G_1}$ = 10.0 hr, σ_{G_1} = 2.0 hr; $\bar{\tau}_s$ = 20.0 hr, σ_s = 4.0 hr; $\bar{\tau}_M$ = 2.0 hr, σ_M = 0.8 hr.

Thus, the model simulated the organization of proliferative processes in the plateau phase of the growth of a LICH cell culture different from the one described in publications [7, 33]. The principal difference is that the stationary culture is assumed to consist by 70 percent of cells resting outside the cycle (true G_0 phase) and by 30 percent of proliferating cells whose mitotic cycle duration is greater than that of the cells in an exponentially growing culture. In the simulation experiments described below the cloning of cells occurred with the temporal parameters of the mitotic cycle peculiar to the exponential state of the population, i.e. with $\bar{\tau}_c$=28 hr and σ_c=3.4 hr. Except for the characteristics of the G_0 phase which describe radioresistance P_{G_0} and the level of repair processes C_{G_0}, all the parameters of the model were identical to those in experiments 1,2 and 3.

Simulation experiment 4. The purpose of the experiment was to estimate the radioresistance P_{G_0} of the resting LICH cells. This can, apparently, be done by choosing such a value for this sole unknown parameter which would ensure a reasonable agreement between the simulation results and experimental evidence. The parameter was adjusted by the survival rate value for LICH cells when cloned immediately after irradiation with a dose of 5 Gy. The value of P_{G_0} determined in this way proved to be equal to 0.79. Inasmuch as a dose of 5 Gy contains 20 elementary doses, it may be concluded that LICH cells in the plateau phase of growth are considerably more sensitive to the damaging effect of radiation than those on the phase of exponential growth.

Simulation experiment 5. Using the survival value for cells irradiated with a dose of 5 Gy and subcultured 6 hours after irradiation [34] , the C_{G_0} value may be estimated.

Indeed, with this set-up of the experiment the value C_{G_0} may be unambiguously identified by experimental data on the survival of cells subcultured 6 hours after irradiation since, with a known P_{G_0}, these data contain information only on the repair processes occurring in the resting cells. The obtained estimator C_{G_0} = 0.25 indicates that the processes of the repair of damage caused by a dose of 5 Gy are almost twice as active in the resting cells as they are in the proliferating state.

Hence, on the basis of the results of simulation experiments 4 and 5 the following important conclusion may be drawn: the resting cells may simultaneously possess both a higher sensitivity to radiation and a greater capacity for postradiation repair than their actively proliferating counterparts.

It should be noted that it is not always at all, as in the case under review, that the survival of cells irradiated in the stationary phase of cell culture growth is below that of cells exposed to radiation in the exponential phase [18].Kim et al. [11] have found that the survival of 3T3 cells irradiated in the stationary phase of growth is much higher than in the case of the radiation treatment of those cells in an exponentially growing culture, whereas the survival rates for HeLa cells irradiated in the two states are practically identical.Only slight differences in the survival of cells irradiated in the stationary and exponential phases of cell culture growth have also been noted in experiments with the recently obtained RIF-1 tumour line [21] . It is worthy of notice that intensive PLDR takes place in the stationary culture of those tumour cells, yet in vivo the clonogenic potential of either small or large-sized tumours does not change with increase in the time interval between irradiation and cloning. Moreover, RIF-1 tumour cells in vivo show a survival ten times as high as in vitro. The possible explanations of these peculiarities of RIF-1 tumour are:

(a) repair processes in the cells of this tumour in vivo do not actually occur due to their inhibition by large doses of radiation, and the high level of survival of these cells is ensured only by the functioning of special mechanisms responsible for cell radioresistance characterized by the parameter P. It is not unlikely either that the mechanisms ensuring radioresistance and functioning at the level of peak activity may also exert an inhibiting effect on the processes of radiation damage repair;

(b) PLD repair processes do occur, but are so rapid in RIF-1 tumour cells in vivo that they cannot be detected within the time intervals that are used;

(c) the characteristic time parameters of the processes of radiation damage formation and repair are of comparable magnitude, and the interaction of those processes is such that the standard set-up of the experiment to reveal PLDR proves inefficient.

In conclusion, it should be pointed out that with other experimental tumours the possibility of PLDR has been very convincingly demonstrated [15] .

Simulation experiment 6 (model testing). In this experiment the adequacy of the model was tested by simulating cell survival determined both immediately and 6 hours after irradiation with doses of 4 and 6 Gy. In this simulation experiment all the parameters of the model were assigned values assessed in the independent experiments described above. The results presented in Figure 3.9 show a satisfactory agreement between the predictions of the model and the experimental data. These results also suggest a conclusion that the repair function - irradiation dose relationship B(D) for the resting cells differs

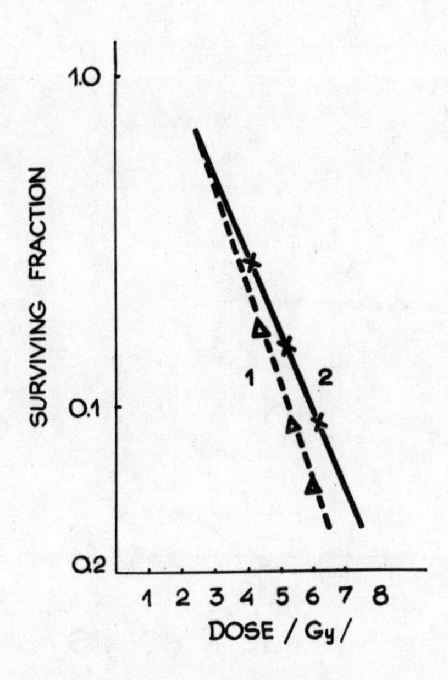

Figure 3.9. Survival of LICH cell culture irradiated in the stationary phase of growth. Subculture 1 - immediately after irradiation, 2 - 6 hours after. Solid lines drawn through experimental points from [34] . Δ and ✕ - mean survival values for 5 realizations on the simulation model.

but little from that for actively proliferating cells. This conclusion may be made on the strength of the fact that, despite the fixed B(D) relationship (identified from the effect-dose curve for an exponentially growing cell population), a satisfactory agreement is attained between the results of simulation and biological experiments.

Simulation experiment 7. The aim of this experiment was to reproduce the kinetics of potentially lethal damage repair of LICH cell culture. The results of the experiment for a dose of 5 Gy are shown in Fig.3.10. They indicate that the survival of cells irradiated in the stationary phase of growth, unlike those of an exponentially growing population, may be enhanced if the explantation of the cells to determine their clonogenic capacity is delayed for a certain time $\Delta t_{clon.}$ after irradiation. The course of survival variations in time depicted in Fig.3.10 corresponds to that observed in the actual experiment [15] . As seen from Figure 3.10 PLDR of LICH cells practically terminates 6 hours after irradiation. Within the interval of 6

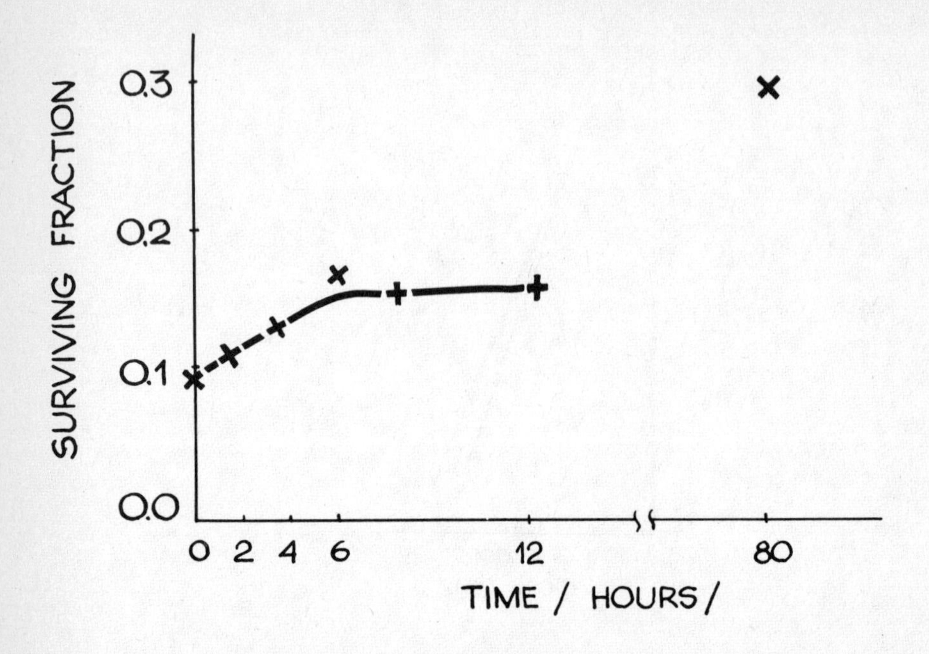

Figure 3.10. Relationship between survival of cell culture irradiation with a dose of 5 Gy and the time interval between irradiation and explantation, × - mean values for six realizations.

to 12 hours after irradiation the survival variation curve reaches plateau. Further considerable rise in survival, recorded 80 hours after irradiation, is due not to PLDR proper but to the postradiation selection of cells owing to the reproductive (and interphase) death of the unviable elements of the cell population. The role of selection leading to the enrichment of the population with viable cells is often underestimated in interpreting PLD and SLD repair [28, 29, 31, 32]. Thus, the dominant part in the effect of enhanced survival (PLD repair) observed with increase in $\Delta t_{clon.}$ is played by cells in the true G_0 phase of the mitotic cycle, whereas prolongation of the mitotic cycle in a stationary cell culture cannot serve as a mechanism responsible for that effect. This conclusion is also supported by the results of a simulation experiment designed to study the kinetics of PLDR in a population containing 100 percent of resting cells (see simulation experiment 12).

The PLDR effect noted in the exponentially growing culture of

RIF-1 tumour cells [21] may be accounted for by the presence in the culture of a certain fraction of resting cells, the authors not attempting to detect it. Thus, the results of this simulation experiment may be justly regarded as a radiobiological proof of the presence of resting cells outside the mitotic cycle in the plateau phase of the growth of LICH cell culture. They form the basis for the assumption of the presence of such cells in a RIF-1 tumour exponentially growing in vitro. As pointed out above, the existence of 70 percent of resting cells in a stationary LICH culture is corroborated by analysis of the kinetics of cell populations in that culture [6]. The foregoing does not imply that proliferating cells are incapable of undergoing PLDR. Thus, in [6] arguments are adduced for the possession of such capacity by cells of a stationary LICH culture in progress through the MC. It is evident from Figure 3.9 that with the conventional set-up of the experiment on revealing the effect of PLDR of cells there is a change in the slope of the effect-dose curve. This phenomenon was observed in the majority of radiobiological investigations using tumour cells. It would be premature, however, to regard it as universal since some authors (in experiments with normal cells cloned in vivo) recorded the PLDR effect manifesting itself only in a longer shoulder of the survival curve, without any substantial change in its slope [5].

In the past few years considerable progress has been made in elucidating the role of PLDR in the radiotherapy of malignant neoplasms. Investigations using nine lines of human tumour cells cultivated in vitro [28-32] have shown that in cells obtained from radiocurable tumours PLDR in a stationary culture is much less pronounced than in those from radioincurable neoplasms. Once of the spectacular results obtained by the authors employing that approach is the conclusion that intensive PLDR by resting tumour cells in the interval between consecutive irradiation fractions may account for the failure to ensure local control of a radioincurable tumour with certain regimes of fractionated treatment. It has been earlier shown that tumour cells irradiated in hypoxia (both acute and chronic) recover more effectively from PLD than those exposed to irradiation under the conditions of normoxia [25].

In this connection, however, reference should be made once again to the results of studying the RIF-1 tumour which vividly demonstrate significant differences in the response of tumour cell populations to irradiation in vitro and in vivo.

In one of his latest works Weichselbaum [27] showed the existence in human tumours of cell clones possessing a distinctly elevated capacity to undergo radiation damage repair. This is a heritable property of a clone, not one determined only by the conditions of cells' microenvironment, as it was formerly believed. Further experimental and clinical studies are required to establish conclusively the contribution of PLDR to the radioresistance of specific human tumours and to ascertain the prospects of utilizing the knowledge obtained on the PLDR kinetics in the search for ways to enhance the efficiency of radiotherapy. In our opinion, simulation of the spectrum of possible tumour responses to single and fractionated irradiation may contribute greatly to the success of such studies. Simulation of the effect of fractionated irradiation will be the subject of the next section.

Simulation experiment 8. This experiment was staged to ascertain the relationship between PLDR kinetics and the degree of radiation injury. The experimental conditions were similar to those of experiment 7, except for the dose of irradiation. The results presented in Figure 3.11 show that increase in irradiation dose from 2 to 6 Gy is ac-

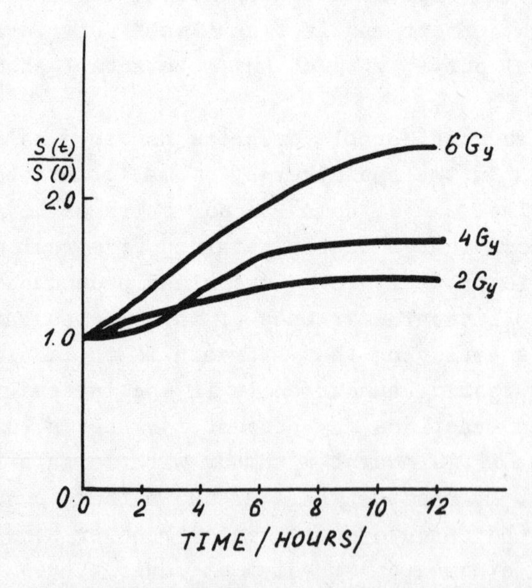

Figure 3.11. Relative increase in cell survival with increasing explantation time (Δt_{clon}) for a cell culture in the stationary phase of growth after irradiation with a dose of 2 to 6 Gy (mean values for 5 realizations of simulation experiments).

companied by a rise of the upper survival level attainable through delayed explantation of cells irradiated in the stationary phase of growth. This effect was also observed in a real experiment [14] . It would be premature to associate it with a possible activation of the repair system coincident with increase in irradiation dose since this assumption was not adopted in the model. The trend of the curves in Figure 3.11 is adequately accounted for by the non-linear contribution to the survival of cells by their redistribution among damage levels both immediately after irradiation with different doses and in the course of radiation damage repair.

3.5. Effects of Fractionated Irradiation

All the results of simulation experiments presented below were obtained with the same values of the parameters of the model as were used in the analysis of single exposure effects. In these experiments variations in the survival of LICH cells were studied in the case of fractionated irradiation of a cell culture in the exponential and stationary phases of growth. To facilitate comparison with the results described above the 5 Gy dose was divided into two equal (2.5 Gy) fractions and the accumulated effect of the two fractions was studied with different time intervals allowed between two successive irradiations. The program for the simulation of fractionated irradiation effects has been described in Section 2.9.

Simulation experiment 9. The conditions of the experiment were as follows: an exponentially growing culture of LICH cells, the 28-hour mean duration of the mitotic cycle, and the "expiration time" of 50 hours. The curve constructed by means of simulation and representing cell survival as a function of time interval Δt between two fractions is shown in Figure 3.12 (curve 1). That character of cell survival curve in the event of fractionated irradiation is quite often observed in actual practice [18] . Unfortunately, we have failed to find in the literature corresponding experimental evidence for cells of the LICH line. However, survival variation curves following fractionated irradiation obtained on numerous lines (primarily, of tumour cells) have no typical extremum with short time intervals between fractions, being monotonically increasing curves. It is reasonable to assume that the type of curve is dependent substantially on the value of the "expiration time". The following experiment was undertaken to test this assumption.

Simulation experiment 10. The experimental conditions were similar

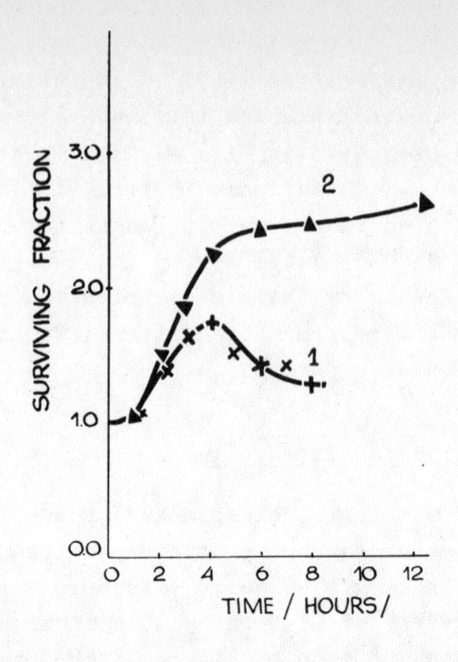

Figure 3.12. Results of simulation experiments in fractionated irradiation (2.5 Gy + 2.5 Gy) of a LICH cell culture in the exponential phase of growth at different time intervals.
Curve 1 - "expiration time" of 50 hrs.
Curve 2 - "expiration time" of 7 hrs.
Mean values for three realizations are presented.

to those of experiment 9, except that the value of the "expiration time" was reduced to 7 hours. The survival variation curve obtained at that value of the "expiration time" after fractionated irradiation is also displayed in Figure 3.12 (curve 2). It is evident from Figure 3.12 that both the value of relative increase in cell survival after fractionated irradiation and the form of the curve which reflects the kinetics of sublethal damage repair are dependent in a large measure on the "expiration time".

Simulation experiment 11. The purpose of this experiment was to study the dynamics of cell survival in a stationary culture after fractionated irradiation. The set-up of the experiment was similar to that used by Little et al. [15] in studying the additive effect of sublethal damage and potentially lethal damage repair in a stationary culture of LICH cells. It provided for irradiation with a dose of 2.5 Gy, at the instant t=0, reirradiation with the same dose at different time intervals Δt and explantation of cells to determine their clonogenic capacity right after reirradiation. The parameters of the model were identical with those used for the plateau phase of growth of LICH cells (Section 3.4). The results of the simulation experiment are shown in Figure 3.13: curve 1 corresponds to the "expiration time" equal to 50 hours, and curve 2 to that of 7 hours.

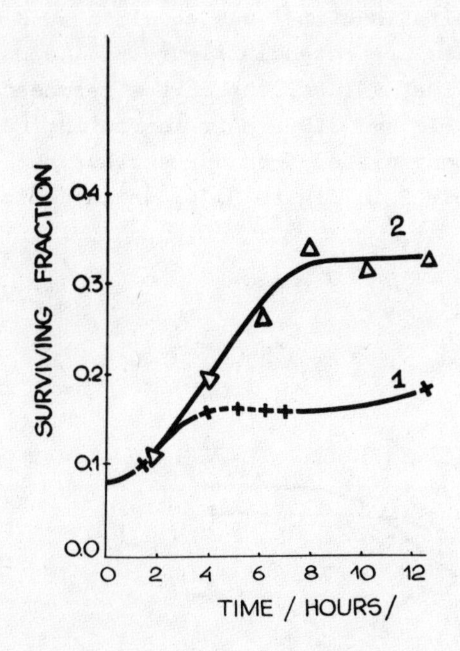

Figure 3.13. Variations in survival of LICH cells in the stationary phase following fractionated (2.5 Gy + 2.5 Gy) irradiation at different time intervals for "expiration time" values of 50 hrs - curve 1 and 7 hrs - curve 2. Mean values for three realizations are presented.

The run of curve 2 in Figure 3.13 indicates that with the selected set-up of the experiment and a small value of the "expiration time" variations in the survival of cells after fractionated irradiation are contributed to appreciably by the processes of both SLDR and PLDR. This result is in fundamental agreement with the data by Little et al. [15] who used in their experiments a somewhat different irradiation regime, with the first fraction being equal to 5 Gy and the second to 3.5 Gy. It also attests that LICH cells in the stationary phase of growth are characterized not only by a high ability for PLD repair but also by a comparatively short "expiration time" which is the main cause of extra enhancement in survival in the event of irradiation with a fractionated dose.

Simulation experiment 12. Inasmuch as with an "expiration time" of 50 hours the kinetics of PLD repair differs but little from that of SLD repair (see Figures 3.10 and 3.13) it was of some interest to simulate these processes for an idealized population consisting by 100 percent of resting (G_0) cells. The other conditions of this experiment were similar to those of the previous experiments. The chosen value of the "expiration time" was equal to 50 hours. The results of the experiments presented in Figure 3.14 demonstrate that, with large "expiration time" values, the curves representing respectively the processes of PLD and SLD repair in resting cells (curves 1 and 2 in Figure 3.14) may differ from one another but very slightly. The behaviour of curve 2 in Figure 3.14, which depicts the ki-

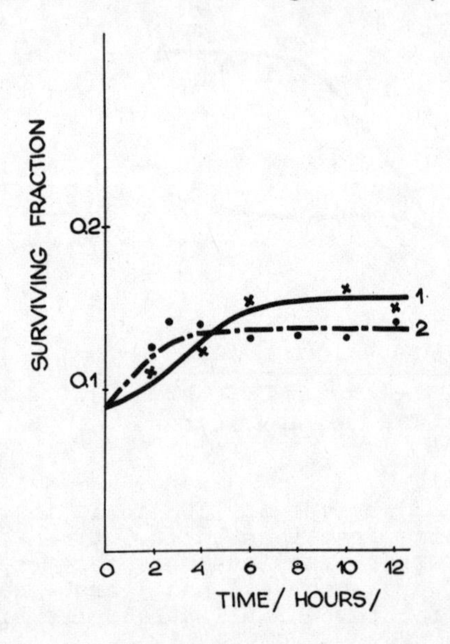

Figure 3.14. Variation in survival of a culture consisting by 100 per cent of resting (G_0) cells after irradiation with a dose of 5 Gy and subculturing at different time intervals - curve 1, and after irradiation with two fractions of 2.5 Gy at different time intervals between successive irradiations, with subculturing immediately after the second exposure - curve 2.
Mean values are given for six realizations.

netics of PLD repair in a population consisting exclusively of resting cells, corroborates the foregoing conclusion that it is cells in the G_0 phase that it presumably responsible for variations in the survival of cells in the plateau phase with varying time intervals between irradiation and cloning.

3.6. Factorial Variance Analysis of Simulation Results

Present-day cell radiobiology is noted for detailed studies of different factors capable of a substantial contribution to ultimate radiation effect recorded by determining the clonogenic capacity of cells [2, 19] . Considering the simultaneous action of numerous factors, quantitative evaluation of the contribution of each to cell population response to irradiation presents an arduous task. Clearly, a purely empirical approach to resolving the problem cannot be effective (due to the unobservability of many factors and scarcely feasible requirements for a body of experimental evidence), and a constructive alternative is investigation into the properties of models adequately representing all known peculiarities of the behaviour of irradiated cell populations.

Methods of direct computer-assisted simulation of biologic processes lead to the development of highly realistic models.Such models, however, are invariably multiparametric, and investigation of their own properties also entails considerable technical difficulties. In other words, an intricate simulation model is becoming the subject of inquiry in itself calling for specialized software.

In this section of the book we shall endeavour to assess by means of simulation modelling the part played by certain factors in the response of a cell culture in a stationary phase growth to single irradiation. In line with Shannon's recommendations [23] we use for this purpose factorial variance analysis of the results of simulations. The fundamentals of variance analysis are expounded in numerous manuals, e.g. in [1, 9, 22] .

By variance analysis is meant a statistical method for processing the results of observations depending upon different simultaneously operating factors. Its aim is to evaluate the contributions of such factors and their interactions to the variation of a certain output quantity presumably dependent on them. Evaluation of this effect is performed with a certain prescribed probability. Analysis of variance is employed in studying factors of a qualitative nature or quantitative factors treated as qualitative ones. Fundamentals of this method were developed by R.Fisher. His idea consists in resolving the total variance of a random variable into independent random components conditioned by the influence of independent factors and their interactions and residual variance associated with random effects (experimental error) unknown to the experimentor.

In order to decide whether the effect of some factor is or is not significant it is necessary to evaluate the significance of a variance component conditioned by the given factor in relation to the variance conditioned by experimental error. The significance of variance evaluations is verified with the F-ratio test. To ascertain which particular factor levels differ from each other the significance of differences between the levels is evaluated.

A variance analysis model represents each observation in the form of a linear combination of the common mean value, factor effects, their combinations and a random error. For instance, in the case of two factors each observation is represented as follows

$$y_{ijk} = \mu + \alpha_i + \beta_j + \gamma_{ij} + \varepsilon_{ijk} ,$$

where y_{ijk} is the kth repetition at the ith level of factor 1 and at the jth level of factor 2; μ is the common mean value; α_i is the ith level effect of factor 1; β_j is the jth level effect of factor 2; γ_{ij} is the effect of interaction of the ith level of factor 1 and the jth level of factor 2; and ε_{ijk} is the random error in the given experiment. The following restrictions are introduced for ε_{ijk}: the errors are assumed to be independent and normally distributed with zero expectation and equal variances.

To test hypotheses of the significance of factor - or factor interaction effects corresponding F-ratios are constructed for comparing factor-conditioned variations with that resulting from a random error, allowing for corresponding degrees of freedom. The values obtained for the F-ratios are compared with tabulated values. If any one of them is greater than the one in the table, it means there exists a difference in factor effects. Application of the F-ratio test makes possible only a general conclusion of differences between factor level effects. If the hypothesis of the equality of all factor effects is rejected, the investigator will be interested to know the effects of which particular factor levels are responsible for rejecting the hypothesis. For that purpose the methods of multiple factor comparison are used.

To find out the factors contributing decisively to the survival of cells exposed to a single irradiation we simulated the effects of irradiating a stationary culture with doses of 2 and 5 Gy and analyzed the group of controlled model parameters (Table 3.2).

Symbols $\bar{P}_{MC}$ and $\bar{C}_{MC}$ designate parameters P(W) and C(W) averaged over the mitotic cycle. The factor "variability of MC phase durations" was simulated by multiplying the prescribed values of phase duration

Table 3.2

The values of factor levels for complete factorial design

Factor	Level No.			
	1	1	3	4
1. Radiosensitivity P and repair efficiency C values (for G_o phase relative to MC)	$P_{G_o}=\bar{P}_{MC}=0.84$ $C_{G_o}=\bar{C}_{MC}=0.14$	$P_{G_o}=0.79$ $\bar{P}_{MC}=0.84$ $C_{G_o}=0.25$ $\bar{C}_{MC}=0.14$	$P_{G_o}=0.79$ $\bar{P}_{MC}=0.84$ $C_{G_o}=0.05$ $\bar{C}_{MC}=0.14$	$P_{G_o}=0.89$ $\bar{P}_{MC}=0.84$ $C_{G_o}=0.25$ $\bar{C}_{MC}=0.14$
Dose, Gy	2	5	-	-
Coefficient at MC phase duration variance	1	3	5	-
Proportion of G_o cells in the total number of cells, %	10	50	90	-
2. Mean MC duration, hr	16	32	48	-

variances by the factors 1,3 and 5. The experiments conducted according to full factorial design of the type $3^3 \cdot 2$ (each design including two replications) yielded eight tables of results. Their analysis led to conclusions of the significance of the factors under study. As one would expect, the contribution of the "dose" factor is decisive in any combination of factors. However, the significance of other factors may vary with the irradiation dose and therefore it is appropriate to classify the "dose" factor among the controlled parameters determining the conditions of a simulation experiment. The significance of the factors "mean MC duration", "variability of MC duration" and "fraction of resting cells (G_0 cells)" as well as of interactions of these factors, was studied separately for four variants of radioresistance P and repair efficiency C relationships ratios between resting and proliferating cells: 1) $P_{G_0} = \bar{P}_{MC}$, $C_{G_0} = \bar{C}_{MC}$; 2) $P_{G_0} < \bar{P}_{MC}$, $C_{G_0} > \bar{C}_{MC}$; 3) $P_{G_0} < \bar{P}_{MC}$, $C_{G_0} < \bar{C}_{MC}$; 4) $P_{G_0} > \bar{P}_{MC}$, $C_{G_0} > \bar{C}_{MC}$. It should be noted that analysis of experimental data from studying thesurvival of irradiated LICH cells performed in simulation experiment 5 has indicated that the realistic ratios are represented by variant 2. However, the universaltiy of such ratios (variant 2) between the parameters P_{G_0}, C_{G_0} and $\bar{P}_{MC}$, $\bar{C}_{MC}$ for all cell lines has not been established and therefore other possible ratios were regarded in this work as being on a par with them.

In all the variants, with the exception of variant 4 the principal significant factor (significance level = 0.01) was the fraction of resting cells. In variant 4 which corresponds to irradiation with a dose of 5 Gy and the conditions of $P_{G_0} > \bar{P}_{MC}$, $C_{G_0} > \bar{C}_{MC}$ all the investigated factors were insignificant. Thus, it may be presumed that decline in the significance of different characteristics of cell kinetics for the survival of cells with increase in irradiation dose may be largely due to the presence of a cell fraction highly resistant to radiation injury. It will be seen from Table 3.3 that wherever resting cells offer lower radioresistance but higher ability to post-radiation repair than do proliferating cells, as is the case with LICH stationary culture, the factor " mean MC duration " is insignificant. However, for other ratios between resting and proliferating cell characteristics (variant 1, irradiation dose 2 Gy and variant 2, irradiation dose 5 Gy) this factor affects significantly the survival of an irradiated population. The factor "variability of MC" is significant only in cases where resting cells exhibit higher radiosensitivity and lower capacity to undergo repair

Table 3.3

Results of variance analysis

Variants of radiosensitivity to repair efficiency relationships for cells in G_0 and MC	Dose Gy	Significant factors	Level of significance
Variant 1	2	Fraction of cells in G_0	0.01
$P_{G_0} = \bar{P}_{MC}$; $C_{G_0} = \bar{C}_{MC}$	5	Mean MC duration	0.05
		Fraction of cells in G_0	0.01
Variant 2	2	Ditto	0.01
$P_{G_0} < \bar{P}_{MC}$; $C_{G_0} > \bar{C}_{MC}$	5	Ditto	0.01
Variant 3	2	Ditto	0.01
	5	Fraction of cells in G_0	0.01
$P_{G_0} < \bar{P}_{MC}$; $C_{G_0} < \bar{C}_{MC}$		Mean MC duration	0.01
		Coefficient at variance	0.01
		Interaction of factors "Fraction of cells in G_0" and "MC duration"	0.01
Variant 4	2	Fraction of cells in G_0	0.01
$P_{G_0} > \bar{P}_{MC}$; $C_{G_0} > \bar{C}_{MC}$	5	The factors concerned are insignificant	-

than do proliferating cells (variant 3, irradiation dose 5 Gy).

Thus, in order of significance the factor "fraction of G_0 cells" rates first within the range of doses under consideration. The significance of different factors depends tangibly upon radiosensitivity to repair efficiency relationships for resting and proliferating cells and upon the irradiation dose. Whenever resting cells combine higher radioresistance with greater capacity to undergo damage repair than proliferating cells, then with a fairly high irradiation dose (5 Gy) such factors as "fraction of G_0 cells", "mean MC duration" and "variability of MC duration" or their interactions play no important part in the effect evaluated by cell survival.

Of considerable interest as the subject of similar investigation are the factors P and C for resting and proliferating cells. Such an investigation, as well as variance analysis of a model for fractionated irradiation, is now in progress at our laboratory.

REFERENCES

1. Ahrens, H. and Läuter, J. Mehredimensionale varianzanalyse, Akademie-Verlag, Berlin, 1981.
2. Barendsen, G.W. Variations in radiation responses among experimental tumors, In: Radiation Biology in Cancer Research, Raven Press, New York, 333-343, 1980.
3. Chang, R.S. Continuous subcultivation of epithelial-like cells from normal human tissues, Proc. Soc. Exper. Biol. and Med., 87, 440--443, 1954.
4. Crane, M.A. and Lemoine, A.J. An introduction to the regenerative method for simulation analysis, Springer-Verlag, Berlin-Heidelberg-New York, 1977.
5. Gould, M.N. and Clifton, K.H. Evidence for an unique in situ component of the repair of radiation damage, Radiat. Res., 77, 149--155, 1979.
6. Gushchin, V.A., Zorin, A.V., Stephanenko, F.A. and Yakovlev, A.Yu. On the interpretation of the recovery of cell from potentially lethal radiation damage in stationary cell culture, Studia Biophys., 107, 195-203, 1985 (In Russian).
7. Hahn, G.M. and Little, J.B. Plateau-phase cultures of mammalian cells: on in vitro model for human cancer, Curr. Topic Radiat. Res., 8, 39-43, 1972.
8. Iglehart, D.L. and Shedler, G.S. Regenerative simulation of response times in networks of queues, Springer-Verlag, Berlin, Heidelberg, 1980.
9. Johnson, N.L. and Leone, F.C. Statistics and experimental design in engineering and physical sciences, vol. II, John Wiley and Sons, New York, 1977.
10. Katkovnik, V.Ya. Non-parametric identification and smoothing of data, Nauka, Moscow, 1985 (In Russian).
11. Kim, J.H., Kim, S.-H., Perez , A.G. and Fried, J. Radiosensitivity of confluent density-inhibited cells, Radiology, 106, 447--449, 1973.
12. Little, J.B. Differential response of rapidly and slowly proliferating human cells to X-irradiation, Radiology, 93, 307-313, 1969.
13. Little, J.B. Repair of sub-lethal and potentially lethal radiation damage in plateau phase cultures of human cells, Nature, 224, 804-806, 1969.
14. Little, J.B. and Hahn, G.M., Life cycle dependence of repair of potentially lethal radiation damage, Int. J. Radiat. Biol., 23, 401-410, 1973.
15. Little, J.B., Hahn, G.M., Frindel, E. and Tubiana, M. Repair of potentially lethal radiation damage in vitro and in vivo, Radiology, 106, 689-694, 1973.
16. Nadaraya, E.A. On estimation of regression, Theor. Probab. Appl., 9, 157-159, 1964 (In Russian).
17. Nadaraya, E.A. Non-parametric estimation of probability density and regression curve, Tbilisi, 1983 (In Russian).
18. Pelevina, I.I., Afanas'ev, G.G. and Gotlib, V.Y. Cell factors in tumour reaction to irradiation and chemotherapy, Nauka, Moscow, 1978 (In Russian).
19. Pelevina, I.I., Saenko, A.S., Gotlib, V.Ya. and Synzynys, B.I. Survival of irradiated mammalian cells and DNA reparation, Energoizdat, Moscow, 1985 (In Russian).
20. Rachev, S.T. On Monge-Kantorovich problem and its application in stochastics. Theor. Probab. Appl., 29, 625-657, 1984.

21. Rasey, J.S. and Nelson, N.J. Discrepancies between patterns of potentially lethal damage repair in the RIF-1 tumor system in vitro and in vivo, Radiat. Res., 93, 157-174, 1983.
22. Scheffe, H. The analysis of variance, John Wiley, New York, 1959.
23. Shannon, R.E. System simulation. The art and science, Prentice--Hall, Inc., Englewood Cliffs, New Jersy, 1975.
24. Sinclair, W.K. Sensitization by hydroxyurea and protection by cysteamin of Chinese hamster cells during the cell cycle, In: Radiation Protection and Sensitization, Tailor and Francis LTD, Lond., 201-210, 1970.
25. Urano, M., Nesumi, N., Ando, K., Koike, S. and Ohnuma, N. Repair of potentially lethal radiation damage in acute and chronically hypoxic tumor cells in vivo, Radiology, 118, 447-451, 1976.
26. Watson, G.S. Smooth regression analysis. Sankhya, ser.A, 26, part 4, 359-372, 1964.
27. Weichselbaum, R.R. Radioresistant and repair proficient cells may determine radiocurability in human tumors, Int. J. Radiat. Oncol. Biol., Phys., 12, 637-639, 1986.
28. Weichselbaum, R.R. and Little, J.B. The differential response of human tumours to fractionated radiation may be due to a post--irradiation repair process, Brit. J. Cancer, 46, 532-537, 1982.
29. Weichselbaum, R.R. and Little,J.B. Repair of potentially lethal X-ray damage and possible applications to clinical radiotherapy, Int. J. Radiat. Oncol. Biol. Phys., 9, 91-96, 1983.
30. Weichselbaum, R.R., Nove, J. and Little, J.B. Radiation response of human tumor cells in vitro, In: Radiation Biology in Cancer Research, Raven Press, New York, 345-351, 1980.
31. Weichselbaum, R.R., Malcolm, A.W. and Little, J.B. Fraction size and the repair of potentially lethal radiation damage in human melanoma cell line, Radiology, 142, 225-227, 1982.
32. Weichselbaum, R.R., Schmit, A. and Little, J.B. Cellular repair factors influencing radiocurability of human malignant tumors, Brit. J. Cancer, 45, 10-16, 1982.
33. Zinninger, G.F. and Little, J.B. Cell cycle kinetics and response to fractionated irradiation in exponential and plateau phase cultures of human cells (abstr.), Radiat. Res., 47, 246-247, 1971.
34. Zinninger, G.F. and Little, J.B. Fractionated radiation response of human cells in stationary and exponential phases of growth, Radiology, 108, 423-428, 1973.

IV. SIMULATION OF CONTROLLED CELL SYSTEMS

4.1. Introduction

Cell population homeostasis of renewing tissues is maintained by the dynamic balance between the processes of proliferation of cells and their subsequent death following the fulfillment by each cell of its specialized tissue function. The cell acquires its tissue function competence as a result of the maturation or differentiation processes associated with the mechanisms of its ageing and death. It is taken for granted that the cell composition of such "organized" tissues is regulated on the principle of negative feedback. Many of the existing simulation models of cell systems e.g.[7, 8, 12, 13] contain formalized descriptions of feedback. Their authors, however, made no attempts to study the concrete role of the mechanisms of tissue homeostasis in ensuring the functioning of the cell system under normal conditions and after exposure to damaging agents. In the works by Gusev and Yakovlev [3, 4, 5] a stochastic simulation model was developed designed for studying specific responses of controlled cell systems to external factors. By means of that model some of the possible ways of simulating tumour growth were explored [4] as well as the reaction of the cell system to hydroxyurea [3] . The model describes the dynamics of populations of stem, committed to differentiation and mature cells, allowing for regulatory interactions between them. In synthesizing the corresponding feedback loops, nonlinear relationships were prescribed between regulation signals and cell subpopulations numbers. The structure and the parameters of the model were specified for describing the "crypt-villus" ("C-V") system, which may be considered as the basic proliferating unit of the mammalian small intestine epithelium. It is presumable, however, that many of the model's properties are also common to other continuously renewing tissues. In this chapter we shall deal with a simulation model of the "C-V" system, developed by the same authors, which is a refined version of that proposed earlier in [3, 4, 5]. The new version contains software for simulating radiation injury and post-irradiation repair of intestinal epithelium. Thus, in addition to providing a way of studying normal tissue homeostasis, the model offers a means of analyzing radiobiologic effects.

4.2. Brief Description of the Model

For most of the renewing tissues cell subpopulations may be classified by capacity for proliferation and by degree of maturity. This classification is most naturally constructed in the event of spatial distinction between the processes of cell proliferation and differentiation. A typical example of a tissue posessing this property is the mammalian intestinal epithelium.

In the model the cell system is represented by a set of three principal cell subpopulations (Figure 4.1):

1. Subpopulation of stem cells - D_1 characterized by capacity for selfreproduction which ensures renewal of all subpopulations of the system.

2. Subpopulation of semi-stem cells - D_2 comprising proliferating precursors of mature cells.

3. Subpopulation of differentiated cells (D_3) no longer capable of division, continuing to differentiate and fulfilling the tissue function until death and elimination from the system.

In the model, corresponding to the process of renewal of cell composition is the process of reproduction of the D_1 cells and their transition with a prescribed probability P_{12} to the subpopulation D_2 and then, after passing through the prescribed number n of mitotic cycles, unconditional movement of the D_2 cells to the subpopulation D_3. The time of a cell's residence in the subpopulation D_3 is determined by the cell death rate r and the prescribed value of the minimum time T_3 required for complete maturation of the D_3 cells. The condition of preserving cell monolayer and the order of cell's movement along the villus were accounted for in the model by organizing the elimination of cells from the system according to the "first come, first served" queue discipline.

The structures of the D_1 and D_2 subpopulations are of the same type; they comprise the mitotic cycle (MC) and the resting state G_0 (reserve of cells). The MC is represented by the sequence of phases G_1, S, G_2 and M whose durations were assumed to be independent random variables. As an approximation to the distribution of each phase duration the model uses truncated normal distribution, its parameters being based on experimental data from [14] . The time of cells' residence in the reserve was not generated by a priori prescribed distribution; it was formed by the ratio between cell flows entering the G_0 state after division and leaving that state for a new MC or bound for differentiation according to a "request". Requests were

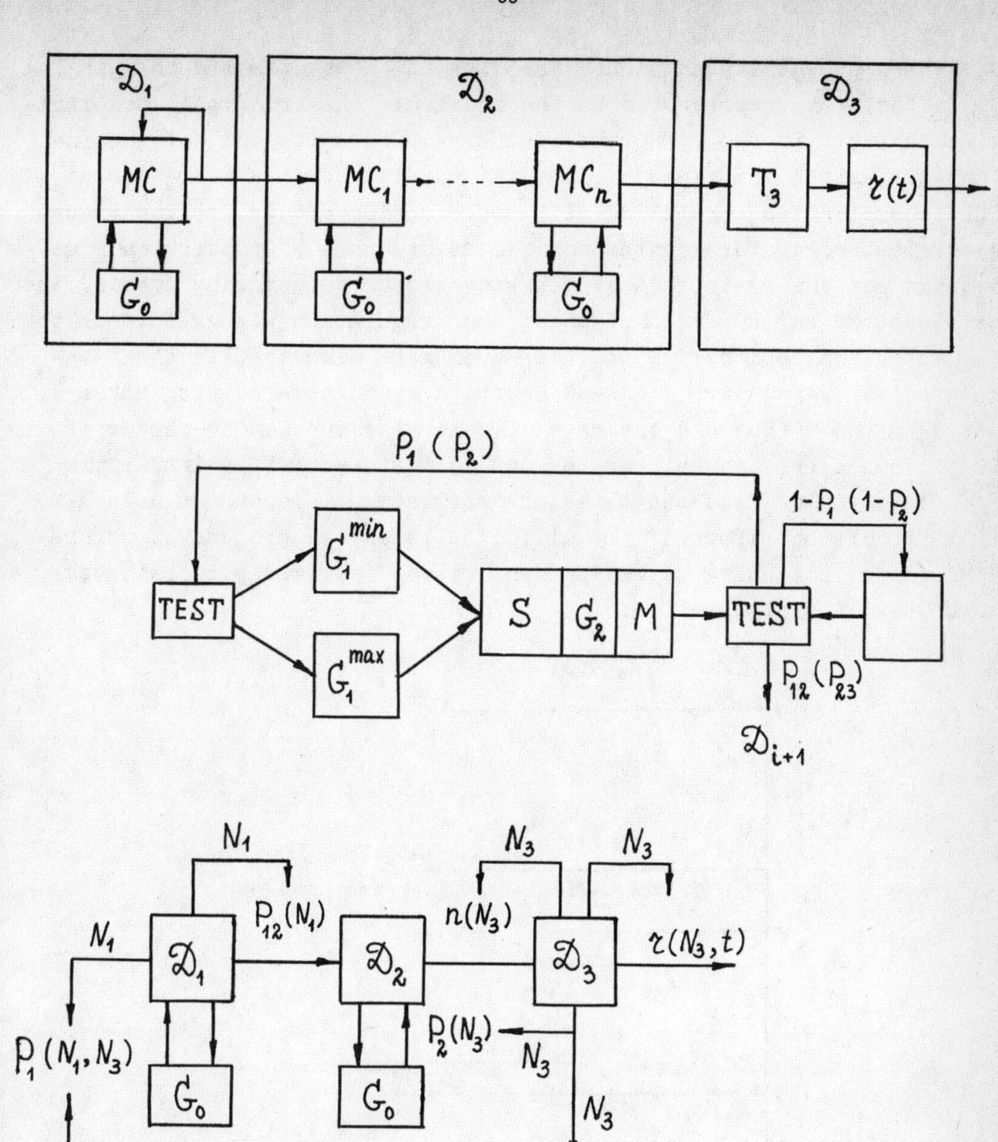

Figure 4.1. Functional block diagram for the model of the "cryptvillus" system.

presented at the instants when the size of the system was changing.

In that way, represented in the model are the principal peculiarities inherent in the structure of the proliferative unit of the intestinal epithelium. In addition, the model describes operation of the tissue system regulating cell reproduction and maturation processes. That regulatory system ensures maintenance of population homeostasis and the resistance of renewing tissue to damaging agents. The existence of various mechanisms of such regulation with reference to the intestinal epithelium was time and again discussed in the literature [15] . Negative feedback has been found between the number of cells on the villus and the rate of cell reproduction in the crypt [2]. Such a relationship may be due to tissue-specific inhibitors of cell division produced by mature enterocytes. Account should also be taken of the effect of the villus cells on the process of maturation which, according to the present views, is rigidly coordinated with that of proliferation.

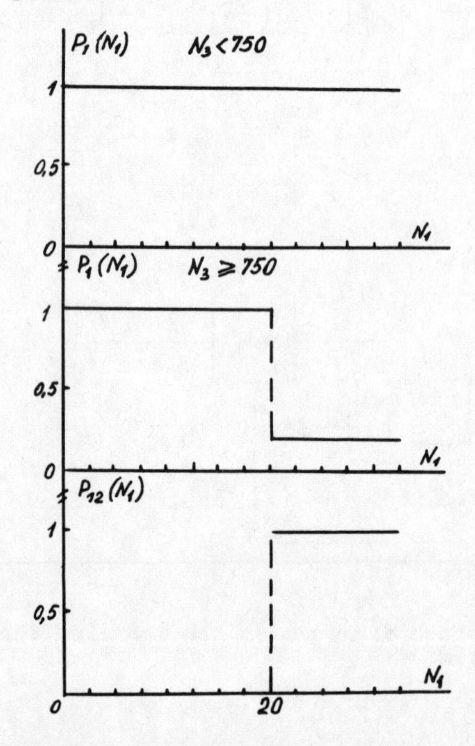

Figure 4.2. Pattern of relationships between stem subpopulation parameters: $P_1(N_1;N_3)$ - probability of entering the MC; $P_{12}(N_1)$ - probability of differentiation.

In accordance with the available data several feedback loops were introduced into the model (Figure 4.1). It was presumed that the probability P_1 of the D_1 cells entering the MC depended on the number of stem cells N_1 and that of mature cells N_3 while the probability P_{12} of the D_1 cells coming for differentiation was dependent

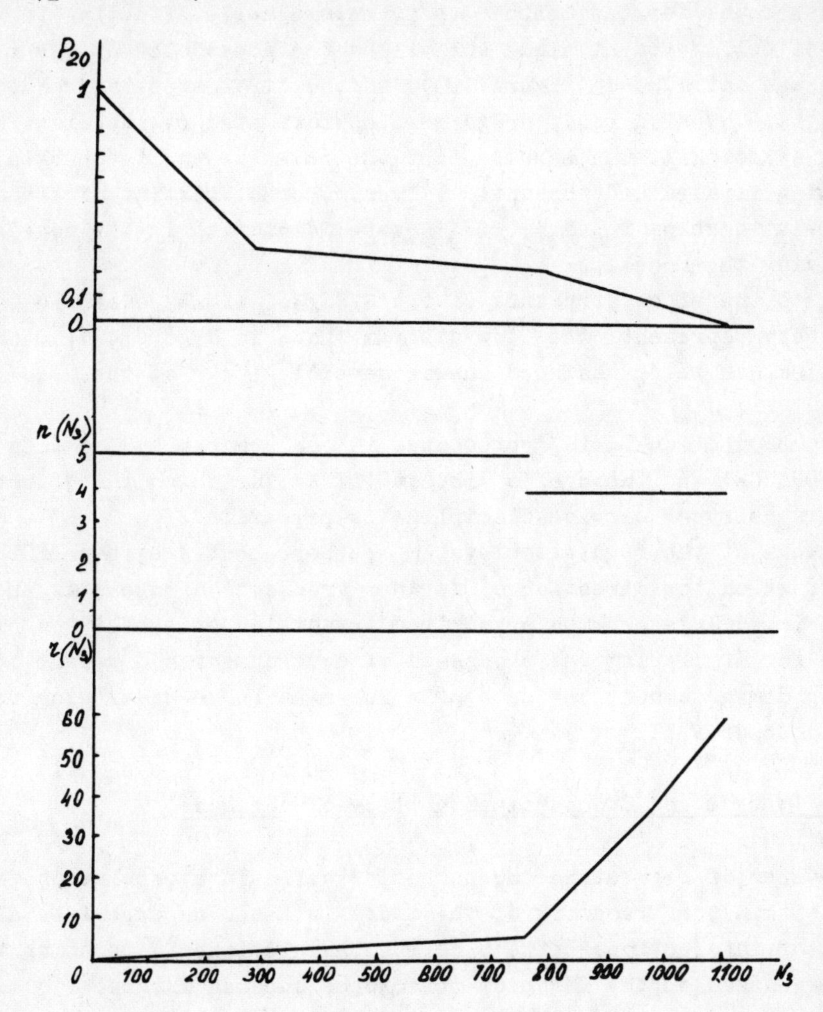

Figure 4.3. Controlled parameters of D_2 and D_3 subpopulations, $P_{20}(N_3)$ - probability of D_2 cells entering the reserve; $n(N_3)$ - number of the MC for D_2 cells; $r(N_3)$ - death rate for D_3 cells.

upon the number of stem cells N_1 (Figure 4.2). It should be noted that the introduced N_1-dependent probabilities follow from the hypothesis of the capacity of stem subpopulation for selfregulation. The probability P_{20} of the D_2 cells entering the reserve depends on the number of mature cells N_3 (Figure 4.3). The number of mitotic cycles for that type of cells was also assumed to be dependent on N_3 (Figure 4.3). This relationship defines changes in cell differentiation rate. To account for the death rate of mature cells affecting the process of cell movement along the villus the dependence of the rate on N_3 was introduced (Figure 4.3). As the first step the relationships $P_1(N_1,N_3)$, P_{12} (N_1), $n(N_3)$ were approximated by stepwise functions. In so doing it was assumed that the intensity of feedback signals in renewing tissues had threshold values. Characteristics of the nonlinear relationships $P_{20}(N_3)$; $r(N_3)$ were identified in the course of utilizing the model.

Owing to the block structure of the GPSS/360 it was possible to exhaustively represent the flow diagram shown in Figure 4.1, each of its elements being assigned one or several blocks of the GPSS/360 system.

The subpopulation D_3 is represented in the program by the delay block (ADVANCE) and the user's circuit (LINK) for which the "first come, first served" service discipline is prescribed.

The links of the regulation system are represented by the TEST blocks in which the direction of further transaction movements are selected in accordance with prescribed transition probabilities. The programs for simulating the processes of development and repair of radiation damage embody the same principles as those underlying the model in Chapter II.

4.3. The Dynamic and Stochastic Stability of the Model

The system of regulating the number of cells in a population formalized within the framework of the model is based on experimental evidence of interactions between cell subpopulations in renewing tissues. Organized on the feedback principle, such an interaction ensures both the steady state of the cell system and its resistance to changes in cell numbers resulting from damaging effects. Synthesis of corresponding controlling units enabled reproduction, in simulation experiments with the model, of these important properties in the case of the intestinal epithelium [3]. Demonstrated in Figure 4.4 is a variant of a simulation experiment on recovery of a model system from a single stem cell. A similar situation may be observed experimentally with the intestinal epithelium exposed to large doses of ionizing radiation. From the results of the simulation experiment it follows that in the proposed model for the regulated cell system of a renewing tissue a steady state sets in which is in the mean sense stable to disturbances of biologic significance (damage and death of cells).

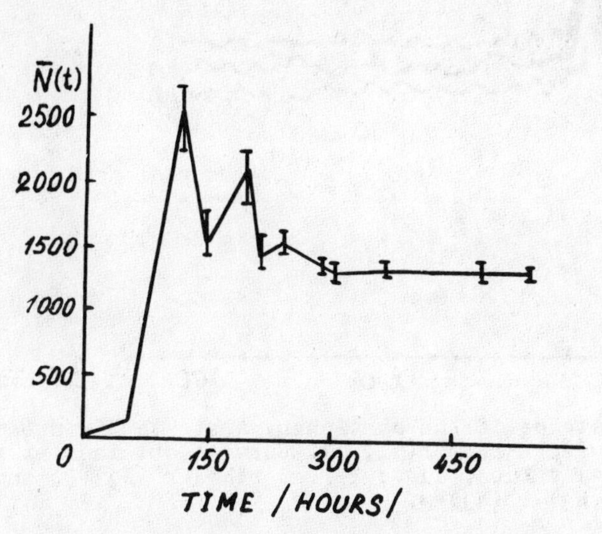

Figure 4.4. Results of a similation experiment on restoring the stationary numbers of the "crypt-villus" system. $\bar{N}(t)$ - the sample mean of the total size (10 realizations). The magnitude of sample standard deviation is shown.

This is also corroborated by another set of simulation experiments in which the prescribed rate of enterocyte extrusion from the villus tips (cell death rate $r(N_3)$) differed from the stationary $r^*(N_3)$:

$$r(N_3) = r^*(N_3) \pm \Delta r ,$$

where Δr is a positive constant.

At $\Delta r \leq 0.1 r^*$ the steady state of the system persisted (Figure 4.5) and the number of cells in the system settled at a level differing but slightly from the initial value. A similar result was obtained with perturbations introduced into the values of other external parameters of the model (MC time parameters for the D_1 and D_2 cells, as well as the number of D_2 cell divisions).

However, the question now arises of whether the control system thus constructed ensures stability in the mean-square sense, i.e. indispensable condition for reliable functioning of a renewing tissue.

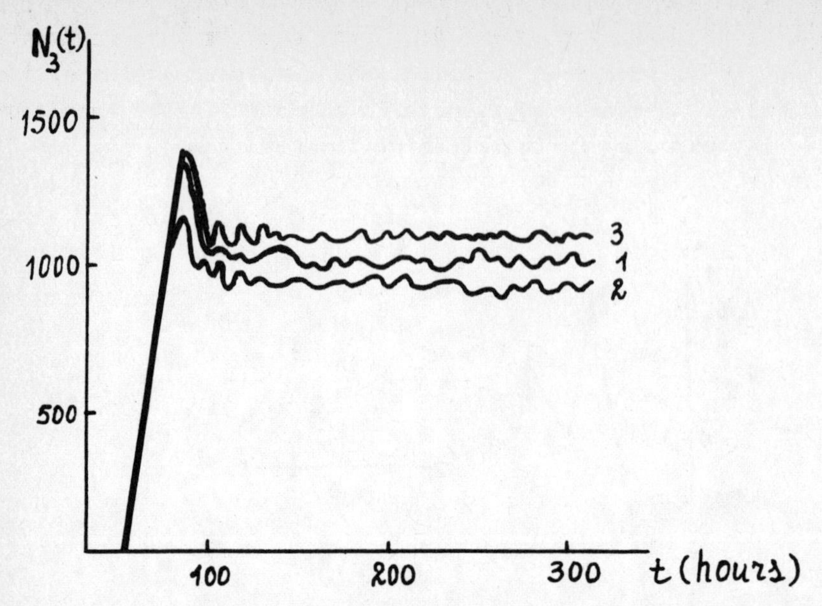

Figure 4.5. Resistance of the stationary state to disturbances in cell death rate $r(N_3, t)$. Curve 1 obtained at stationary $r^*(N_3)$ values, curve 2 - at $r(N_3)=r^*(N_3)+3$, and curve 3 - at $r(N_3)=r^*(N_3)-3$.

It is well known [6] that from the p-stability of dynamic systems at greater p's follows the stability of lesser p's. Moreover, there are cases [11] when the reverse is not true.

By way of example let us consider a linear dynamic system with a negative feedback given by the equation:

$$\dot{X} = (-C+V(t))X \quad , \qquad (1)$$

where $C>0$ is the constant, $V(t)$ is the random noise with the zero expectation and the intensity G^V. Such a system was investigated in [1]. With $U(X)= |X|^p$ chosen as the Lyapunov function the stochastic analogue of the derivative is equal to

$$LU(X) = p|X|^p \left[-C+ \frac{1}{2} G^V + \frac{1}{2} G^V(p-1)\right] .$$

Thus, the conditions for the exponential p-stability of a trivial solution of system (1) (and, consequently, the simple p-stability) are given by the inequality

$$-C + \frac{1}{2} G^V p < 0 \text{ , and hence } C > \frac{1}{2} G^V p .$$

From this it follows that the condition of exponential stability in the mean (p=1) is of the form $C > \frac{1}{2} G^V$, whereas the stability condition in the mean-square sense (p=2) is defined by the inequality $C > G^V$ and appears to be more restrictive.

Analytical study of stability conditions for a model corresponding to the functional scheme in Figure 4.1 presents a more severe problem. On the basis of simulation experiments a rough estimate may be obtained of the output signal variance following introduction of parametric noise into the feedback channels in the steady state of the system. The following sets of simulation experiments were undertaken to investigate several variants of parametric perturbation of the system's steady state. Figure 4.6 displays the dynamics of the sample mean (over 5 realizations) of the total cell numbers in the system. Curve 1 corresponds to the steady state with no random noise.

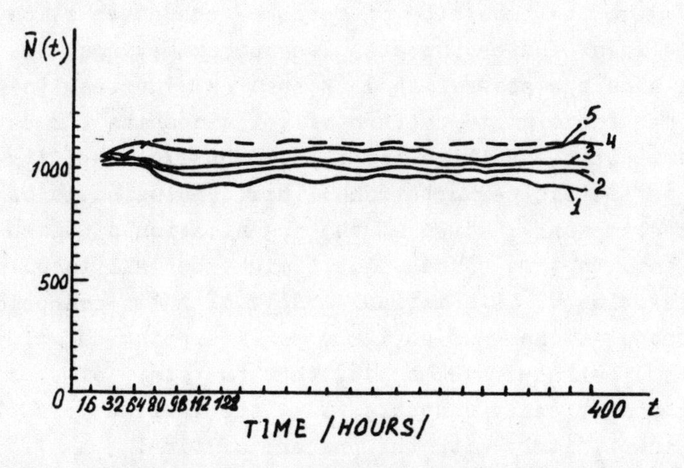

Figure 4.6. Variation in the sample mean value (for 5 realizations) of the total number of cells in the "crypt-villus" system with parametric moises introduced into the feedback channels (explanation in the text).

Curves 2,3 and 4 were obtained following introduction of additive normal noise G(t) with the zero expectation and prescribed standard deviation (σ =100,200 and 300, respectively). The noise was introduced simultaneously into two feedback circuits: for the probability of undergoing differentiation $P_{12}(N_3)$ the argument was given as the sum of the net signal $N_3(t)$ and the normal noise G(t): $P_{12}(N_3(t) + G(t))$; in a similar manner the argument was given for the parameter defining the number of mitotic divisions in a semistem population

$n(N_3(t)+G(t))$. Curve 5 resulted from introducing the additive noise $R(t)$ of greater intensity and distributed uniformly within the interval (0,1000). Analysis of the behaviour of sample mean values suggests the stability, in the mean sense, of the system's steady state to parametric perturbations comparable in quantity to the net (legitimate) signal $N_3(t)$. It is worth noting that curve 5 obtained after the introduction of an additive noise with the peak amplitude for the variants under study is distinguished by the least deviations from the steady state. Thus, it is presumable that parametric noises exert a stabilizing effect and improve the steady state characteristics of a model system. It is known [1] that this is a property of regulation systems of the nonlinear relay type. In our case it appears likely that it is the introduction of stepwise relationships for some regulation parameters that ensures such stabilization of the system under the effect of parametric noises. Futher simulation experiments have shown that with the smoothing of feedback characteristics, i.e. using piecewise-linear instead of stepwise approximations, the controlled system retains the above-listed properties. The results of simulation show that the sigmoid pattern of relationships for parameters to be a sufficient condition for the stochastic stability of a cell system to parametric perturbations. This conclusion is of some interest for experimental studies of the organization of tissue homeostatic mechanisms. In this connection it might be well to point out that the consideration of mathematical models with two independently functioning subpopulations - of actively proliferating and differentiating (maturing) cells - reveals [16] that the state of such a model system is stochastically unstable, i.e. the variance of cell numbers increases indefinitely with the passage of time.

The pattern of changes in the behaviour of the sample variance (Figure 4.7) for the variants of perturbations under review supports the foregoing suggestion. Curves 2,3,4 and 5 in Figure 4.7 differ from curve 1 which represents variations in the sample standard deviation of a population size with no parametric noises. By way of illustration a comparison may be made between maximum (on the observation interval $[0,T]$) values $\max S_i(t)$, $(i=1,2,3,4,5)$, of standard deviations obtained in this simulation experiment: $\max S_1 = 132$, $\max S_2 = 25$, $\max S_3 = 40$, $\max S_4 = 53$, $\max S_5 = 19$.

Based on these results it may be concluded that the presence of parametric noises in feedback circuits may also stabilize the output signal variance, i.e. the number of cells in the system. What is more, a property common to all the curves presented in Figure 4.7 is

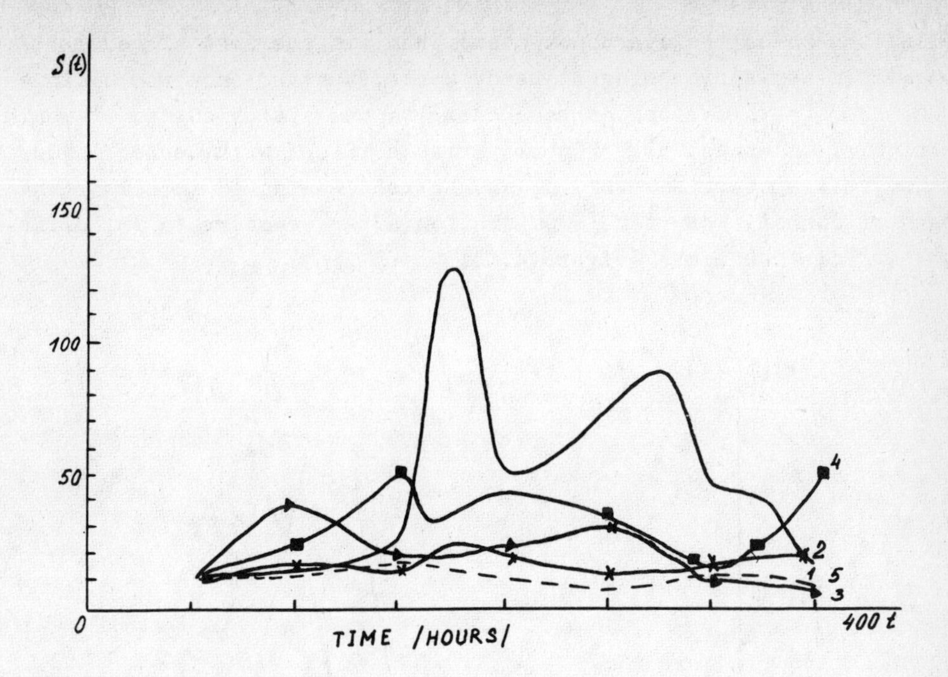

Figure 4.7. Variation in the sample variance of the total number of cells in the "crypt-villus" system with parametric noises introduced into the feedback channels (same variants as in Figure 4.6).

the trend towards the damping of sample standard deviation oscillations which suggests the dynamic stability of the output signal variance to perturbations of the type under consideration.

In the course of investigating the properties of the model the behaviour of the output signal under transient conditions was studied, such conditions developing in the model system when the process of regeneration of cell population after damaging effects was simulated. A typical pattern of model behaviour in the transient state is represented in Figure 4.10b. In a separate set of simulation experiments sample variance values for the total numbers of cells at different stages of the transient process were calculated. In so doing, two qualitatively different stages were distinguished. In the initial stage, with the growing size of the population the variance was increasing, attaining peak values in the period of overproduction of cells (the so-called overshoot peak).Then, in the next stage, as the system was approaching a steady state, the variance was diminishing down to the values corresponding to the steady state. It would be natural to expect higher peak variance values with deeper injury of the cell system. However, no such effect was noted in simulating regeneration of a crypt cell population after exposure to two different irradiation doses (Figure 4.8).

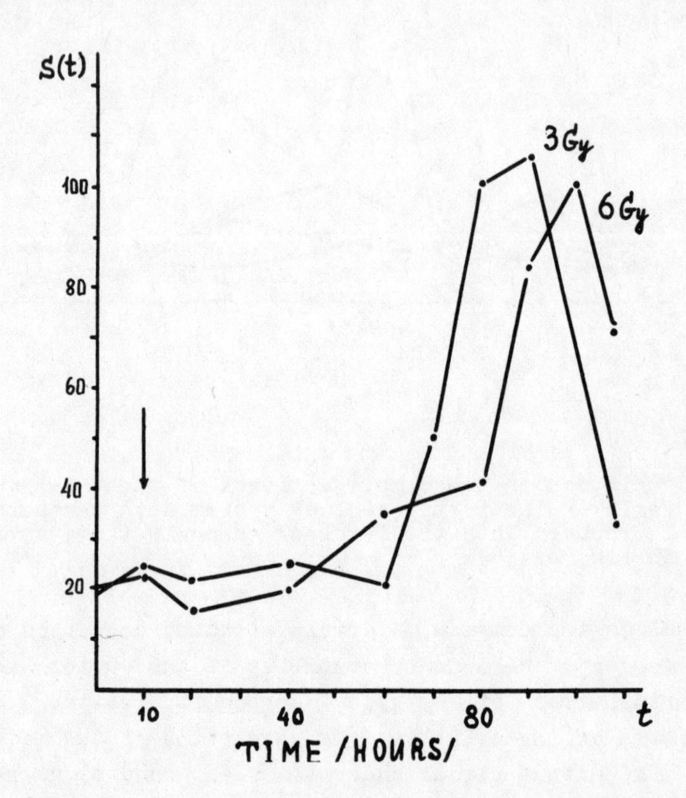

Figure 4.8. Variations in the variance of the total number of cells in the "crypt-villus" system in the transient state after irradiation with doses of 3 and 6 Gy.

Thus, the results of the simulation experiments point to the fact that underlying the maintenance of tissue homeostasis and stability of cell renewal to stochastic perturbations are the common regulatory mechanisms.

4.4. Application of the Simulation Model. Simulation of Radiobiological Effects

The "CV" model was used for the analysis of data on the regeneration of the mouse small intestinal epithelium after acute gamma irradiation with 3 and 6 Gy doses. In conducting simulation experiments use was made of the values of regulation system parameters previously identified in simulating the normal functioning of the C-V system [10]. In reproducing the experiment of irradiating mice with a dose of 6 Gy the values of parameters of program blocks of damage and repair were estimated. In so doing, a satisfactory agreement was obtained simultaneously for two kinetic indices: the number of labelled cells $N_S(t)$ and of mitoses $N_M(t)$ per crypt (Figure 4.9a). Then all the parameters of the model were fixed with the exception of the intracellular repair rate which depends upon the dose and whose variation ensured the agreement between model predictions and experimental data on irradiation of mice with a dose of 3 Gy (Figure 4.9b).

The use of the simulation model made it possible to study the effect of post-irradiation cell repair parameters, unobservable in direct radiobiologic experiment, upon the cell proliferation indices in the "C-V" system. Variation of the repair intensity value in the model did not on the whole affect the shape of the curve. Changes in the repair intensity did affect the value of the peak on the curve of the number of labelled cells observed within 50 to 100 hours after irradiation (Figure 4.9a). Agreement with the experimental curves of DNA-synthesizing cell numbers for a dose of 6 Gy was obtained with a repair intensity 1.5 times lower than that following exposure to an irradiation dose of 3 Gy.

The results of this set of simulation experiments have also made it possible to predict that under certain conditions of irradiation the administration of chemical inhibitors of post-irradiation cell repair may attenuate the end effect of ionizing radiation on intestinal epithelium, i.e. alleviate the intestinal syndrome of radiation disease. This effect caused by mechanisms of tissue regulation was borne out by a direct radiobiologic experiment.

The curves for the size of cell subpopulations obtained from simulation define the structural transformations within the "C-V" system in the early stage of post-irradiation tissue regeneration. Changes in the number of proliferating cells in the crypt and in that of subpopulations of maturing and mature enterocytes are represented sche-

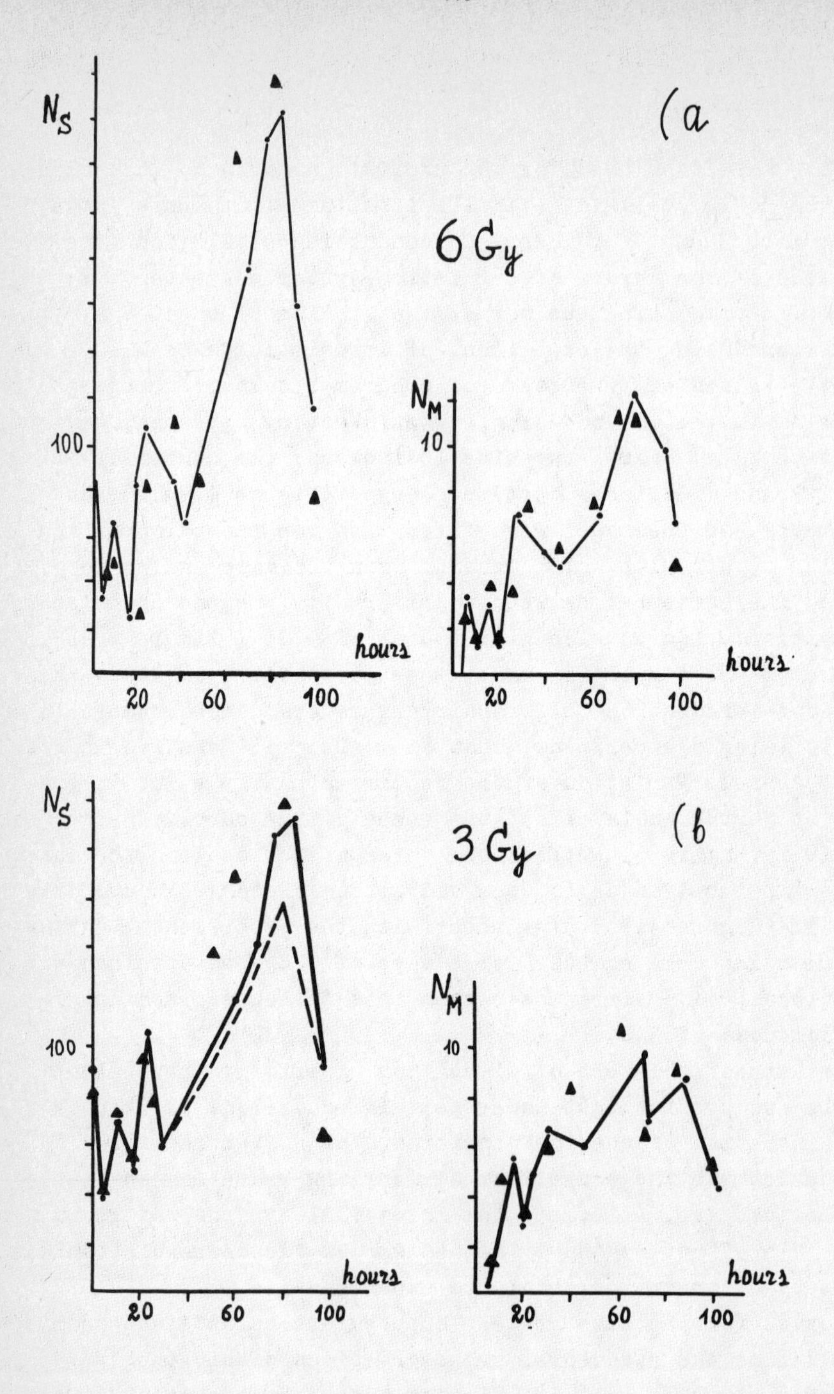

Figure 4.9. Variation in the number of labelled cells (N_s) and of

mitoses (N_M) in the crypts of the mouse intestinal epithelium after a single irradiation with a dose of 3 and 6 Gy. Solid line-simulation results. ▲ - experimental data from [10]. Dotted line indicates changes in the shape of the N_s curve with increasing repair intensity.

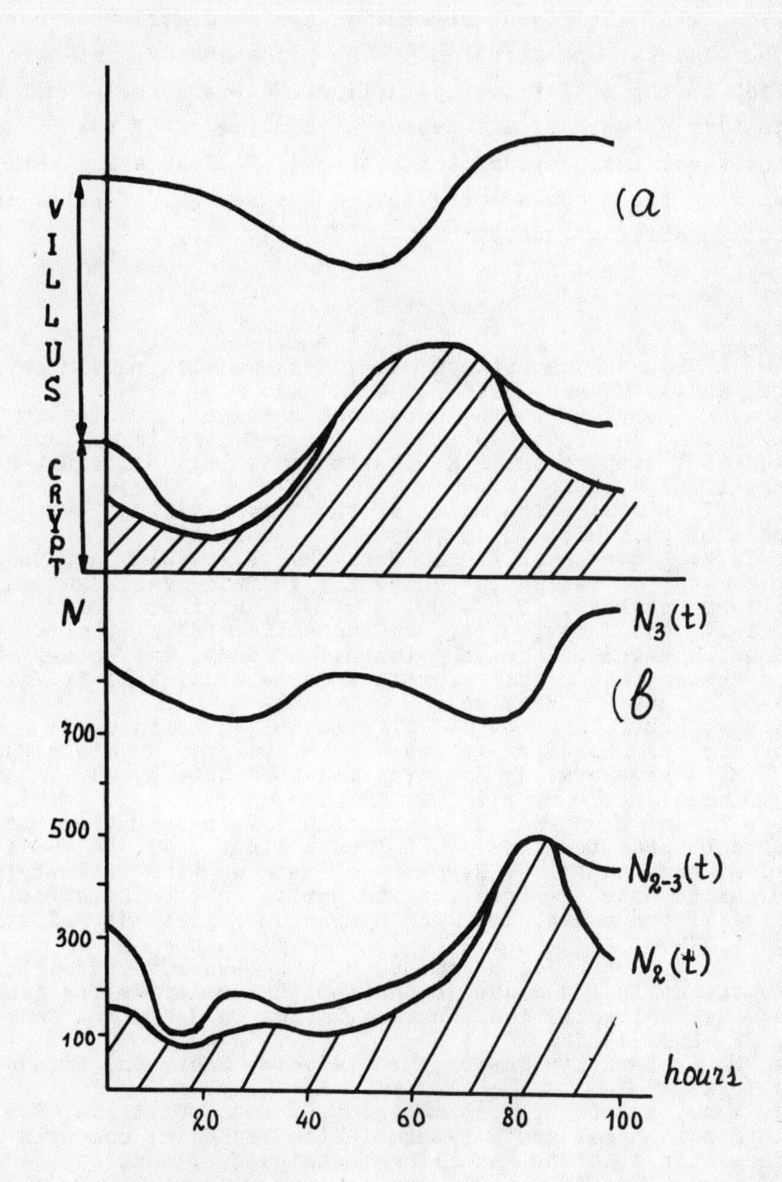

Figure 4.10. Variations in cell subpopulation sizes in the "crypt--villus" system after a single irradiation; a - pattern of variations according to [9] ; b - simulation results.

matically in Figure 4.10b. This pattern of structural transformations obtained by means of the simulation model corresponds in principle to the scheme given in [9] and shown in Figure 4.10a.

From the results of simulation experiments it also follows that the functioning of the tissue regulation system contributes substantially to the temporal organization of the processes of post-irradiation repopulation of the intestinal epithelium. However, the peculiarities of the kinetics of cell proliferation within the "C-V" system do not require for their explanation of hypotheses of changes in the characteristics of the processes regulating the renewal of cells under the effect of ionizing radiation.

REFERENCES

1. Evlanov, L.G. and Konstantinov, V.M. Systems with random parameters, Nauka, Moscow, 1976 (In Russian).
2. Galjaard, H. Some unresolved questions relevant to intestinal adaptation, In. Proc. Intern, Conf. Anat. Physiol. Biochem. Intestinal Adaptation, F.K.Schattauer Verlag, Stuttgart-New York, 1974.
3. Gusev, Yu.V. A simulation model of the "crypt-villus" system, Studia Bioph., 98, 175-182, 1982 (In Russian).
4. Gusev, Yu.V., Zherbin, E.A. and Yakovlev, A.Yu. Simulation of tumor growth initiation in constantly renewing cell systems, Problems Oncology, 30, 74-80, 1984 (In Russian).
5. Gusev, Yu.V. and Yakovlev, A.Yu. Stochastic stability of controlled cell systems. Computer simulation stady, In: Theory of complex systems and methods for their modelling, VNIISI, Moscow, 154-162, 1985 (In Russian).
6. Khasminsky, R.Z. Stability of differential equation systems with randomly perturbed parameters, Nauka, Moscow, 1978 (In Russian).
7. Lamb, J.R. A regenerating computer model of the thymus, Comput. and Biomed. Res., 8, 379-392, 1975.
8. Loefler, T. and Wichmann, H. A comprehensive mathematical model of stem cell proliferation, Cell Tissue Kinet., 13, 543-561, 1980.
9. Lesher, S. and Bauman, J. Recovery of reproductive activity and the maintenance of structural integrity in the intestinal epithelium of the mouse, In: IAEA Symposium Effects of Radiation, Monako, 1968.
10. Lesher, S., Cooper, J., Hagemann, R. and Lesher, J. Proliferative patterns in the mouse jejunal epithelium after fractioned abdominal X-irradiation, Current Topics in Radiation Research, 10, 229-261, 1975.
11. Loeve, M. Probability Theory, 3rd ed., Van Nostrand, Princeton, New York, 1963.
12. Mauer, M.A., Evert, C.F., Lampkin, B.C. and McWilliams, N.B. Cell kinetics in human acute lymphoblastic leukemia: computer simulation with discrete modelling techniques, Blood, 41, 141-154, 1973.
13. Toivonen, H. and Rytömaa, T. Monte Carlo simulation of malignant growth, J.Theor. Biol., 78, 257-267, 1978.
14. Trasher, J.D. The relationship between cell division and cell specialization in the mouse intestinal epitelium, Experientia, 26, 74-76, 1970.

15. Ugolev, A.M. Enterinic (intestinal hormonal) system, Nauka, Leningrad, 1978 (In Russian).
16. Yakovlev, A.Yu. and Zorin, A.V. On simulating the reliability of renewing cell systems, Cytology, 24, 110-113, 1982 (In Russian).

V. THE PROPERTIES OF CELL KINETICS INDICATORS

5.1. Introduction

The problems considered below demonstrate the efficacy of employing computer simulation wherever analytical approaches of the mathematical theory to the investigation of the statistical properties of processes come up against certain technical difficulties. Getting over these difficulties often appears to be the most important stage both in substantiating theoretically the limits of applicability of a particular mathematical model and in elaborating requirements on the accuracy of evaluating from experimental evidence the proposed kinetic indicators.

5.2. Integral Cell Flow into Transitive Population

Let us consider the following formalization of the processes of induced proliferation of cells. Assume that, in response to the proliferative stimulus acting at the instant $t = 0$, a random number m of cells enter the mitotic cycle (and the subsequent cycles) which these cells traverse independently of one another. On completion of the mitotic cycle each cell, independently of the others, produces a random integer number ν of progeny which are immediately recruited into a recurrent division cycle. All the cells have one and the same distribution of the mitotic cycle duration X and one and the same distribution of the progeny numbers ν : the random variables X and ν assumed to be stochastically independent. By introducing the distribution function of the mitotic cycle duration

$$F(t) = P\{X \leq t\} \quad , \quad t \in [0, \infty) , \tag{1}$$

and the generating function of the number of the total progeny

$$h(s) = E\{s^{\nu}\} = \sum_{k=0}^{\infty} P\{\nu = k\} s^k , \ |s| \leq 1 \tag{2}$$

the stochastic age-dependent branching process is defined [2,4,8,9] which is sometimes referred to as the (F, h)-process [9] . The assumption of the independent evolutions of individual cells within a population makes it possible to describe the kinetics of induced cell proliferation by superposition of standard (F, h)-processes [12] .

Let us now distinguish the consecutive phases G_1, S and G_2+M within the mitotic cycle structure and specify the generating function of the number of progeny resulting from the mitotic cell division as follows

$$h(s) = 1 - p + ps^2 , \qquad (3)$$

where (1-p) is the probability of reproductive cell death. This pattern of cell development is represented in Figure 1.1. Let $N_s^c(t)$ denote the stochastic process equal to the number of cells that have entered, by the instant t the S-phase of the mitotic cycle. It will be pointed out that in determining the $N_s^c(t)$ process no particular number of the mitotic cycle is noted, i.e. every cell transition to DNA synthesis contributes to the integral cell flow $N_s^c(t)$.

As shown in the paper by Yanev and Yakovlev [13] , the generating function of the $N_s^c(t)$ process may be derived as follows. Let $\mu_j^i(t)$ be the number of births of the T_j-type cells up to the time t in a process started at t=0 by a single T_i-type precursor cell (i, j=1,2...,n) which was in the population at zero time. Without including the initial cell in the value of $\mu_j^i(t)$, $t>0$, let us introduce the multidimensional generating function of the vector $\mu^i(t)=(\mu_1^i(t),\dots,\mu_n^i(t))$ which is of the form

$$\varphi_i(s;t) = E\left\{s^{\mu^i(t)}\right\} \quad ; \quad \varphi_i(0,t)=1.$$

For the functions $\varphi_i(s;t)$ the following system of nonlinear integral equations is valid [13]

$$\varphi_i(s;t) = 1-F_i(t)+\int_0^t h_i\,[\,s\,\varphi(s;t-u)\,]\,dF_i(u), \qquad (4)$$

where $F_i(t)$ is the distribution of the life span of T_i-type cells, $h(s)=(h_1(s),\dots h_n(s))$ is the multidimensional generating function of the numbers of progeny of each type of cells, and the symbol $s\varphi$ designates the vector $(s_1\varphi_1,\dots s_n\varphi_n)$.

In a special case of the closed structure of the mitotic cycle considered above, the cells in the G_1, S and G_2+M periods will naturally correspond to the types T_1, T_2, T_3. Cells of these types have life-span distribution functions

$$F_1(t) = F_{G_1}(t),\; F_2(t) = F_s(t),\; F_3(t) = F_{G_2+M}(t)$$

respectively. Components of the multidimensional generating functions h(s) for such a process with three types of cells are defined, subject to equation (3), as follows

$$h_1(s) = s_2,\ h_2(s) = s_3\ ,\ h_3(s) = 1-p+ps_1^2\ .$$

In that case system of equations (4) takes the form

$$\begin{aligned}\varphi_1(s;t) &= 1 - F_1(t) + s_2 \int_0^t \varphi_2(s;t-u)dF_1(u)\\ \varphi_2(s;t) &= 1 - F_2(t) + s_3 \int_0^t \varphi_3(s;t-u)dF_3(u)\\ \varphi_3(s;t) &= 1 - pF_3(t) + ps_1^2 \int_0^t \varphi_1^2(s;t-u)dF_3(u).\end{aligned} \qquad (5)$$

What we are interested in is the generating function of the process $\mu_2^1(t)$ i.e.

$$\alpha_1(z;t) = E\left\{ z^{\mu_2^1(t)} \right\} = E\left\{ z^{N_s^c(t)} \right\}\ ,$$

which may be derived from (5), setting $s_2=z$, $s_1=s_2=1$.

As a result we obtain the following integral equation for $\alpha_1(z;t)$

$$\alpha_1(z;t) = 1 - (1-z)F_1(t) - zpG(t) + zp\int_0^t \alpha_1^2(z;t-u)dG(u) \qquad (6)$$

where the convolution $G(t)=F_1(t)*F_2(t)*F_3(t)$ is the distribution function for the entire mitotic cycle duration, i.e. the sum total of durations of the G_1, S and G_2+M phases.

On differentiating (6) with respect to z and setting z=1, we arrive at the equation for the mathematical expectation of the $N_s^c(t)$ process $M(t)=E\left\{N_s^c(t)\right\}$

$$M(t) = F_1(t) + 2p \int_0^t M(t-u)dG(u). \qquad (7)$$

Equation (7) is an equation of a renewal type whose solution presents no difficulties [9] and which is of the form

$$M(t) = \sum_{k=0}^{\infty} (2p)^k F_1 * G^{*k}(t), \qquad (8)$$

where the symbol $G^{*k}(t)$ designates the k-fold convolution of the function G(t).

To generalize expression (8) one may consider the development of a population starting not with one cell, as assumed in deriving (4) and (5), but with m zero-aged cells and treat similarly the superposition of m branching processes. In that case, under the assumption that the mean $\bar{m}$ and variance D_m values of the random variable m are finite, we have the expression for mathematical expectation of the process $N_s^c(t)$ [12]:

$$E\left\{N_s^c(t)\right\} = \bar{m}M(t) = \bar{m}\sum_{k=0}^{\infty}(2p)^k F_{G_1} * G^{*k}(t). \qquad (9)$$

On differentiating (6) twice with respect to z at the point z=1, a little manipulation yields an expression for variance of the process

$$D\left\{N_s^c(t)\right\} = \bar{m}\sum_{k=1}^{\infty}(2p)^k[2M(t)+M^2(t)] * G^{*k}(t)+\bar{m}M(t)+ \\ +(D_m-\bar{m})M^2(t). \qquad (10)$$

Thus, within the axiomatics of a model of a multitype age-dependent branching process,probability characteristics (including mathematical expectation and variance) of a cell flow into the given mitotic cycle phase can be calculated. For a transitive cell population (corresponding to a phase considered apart from other phases of the cell life cycle) it is impracticable to construct the space of elementary events generated by all possible cell evolutions and to define the probability measure for that space.

For a transitive population the probabilistic structure of a random cell inflow may be arbitrary, inasmuch as in that case these is no information on the full evolution of a cell within the cell system under study.

For this reason, in simulating the dynamics of a transitive cell population, one has to confine oneself to the consideration of only mathematical expectations of corresponding stochastic processes. The basic equation for the transient kinetics of a transitive cell population is of the form [14]

$$E\left\{N(t)\right\} = q(t) - \int_0^t g(t-\tau)dF(\tau)+E\left\{N(0)\right\}\int_0^t \frac{1-F(t+\tau)}{1-F(\tau)}W(\tau,0)d\tau , \qquad (11)$$

where $q(t)=\int_0^t E\left\{K_1(\tau)\right\} d\tau$

$K_1(t)$ - the rate of cell entry into a given population;

$F(t)$ - distribution function for the time of a cell's sojourn in the population;

$W(a,0)$ - the density of distribution of a cell age in a given population at the time $t=0$;

$N(t)$ - the number of cells in the population.

The $q(t)$ function was denoted as the q-index of a transitive cell population or an isolated phase of the cell life cycle [14, 15]. So, the q-index of a cell cycle phase is the mathematical expectation of the integral flow of cells into the phase, thus being an analogue of the $M(t)$ function which appears in the description of a closed cell system based on a branching stochastic process model. It should be noted that for processing experimental data the relative q-index ($q^o(t)$) appears to be more suitable, characterizing, as it does, the ratio of the integral inflow $q(t)$ to the initial mean size of the entire cell system, i.e.

$$q^o(t) = \int_0^t E\left\{K_1(\tau)d\tau\right\} / E\left\{N_p(0)\right\} .$$

Here $N_p(0)$ is the total initial number of cells in all the phases of the cell life cycle.

A series of studies [10, 11, 14, 15] have given evidence of the efficacy of employing the q-index of the S-phase of the mitotic cycle in the analysis of the kinetics of induced cell proliferation on the basis of experimental data obtained by means of radioautography techniques. Those studies, however, did not aim to investigate the accuracy of estimating the q-index from temporal patterns of labelling indices. The use of the sampling method for assessing the characteristics, e.g. the correlation function, of the process $N(t)$, calls for such a body of statistical information which cannot be provided by the present-day means of cytological experiment. To overcome the difficulties involved in studying the accuracy and robustness of the procedure for constructing the q-index from experimental data, one will do well to resort to simulation techniques.

Let us direct our attention to the block diagram presented in Figure 5.1 and consider the S phase of the mitotic cycle in isolation, assuming that we do not know the rest of the quantitative interrelations between blocks in the diagram. It will be sufficient for our purposes to restrict ourselves to the case of systems with a low initial level of proliferative activity and assume that the number of cells in the S phase at the instant $t=0$ is equal to zero. In that case equation (11) takes the form

$$E\left\{N(t)\right\} = q(t) - \int_0^t q(t-\tau)dF(\tau) , \qquad (12)$$

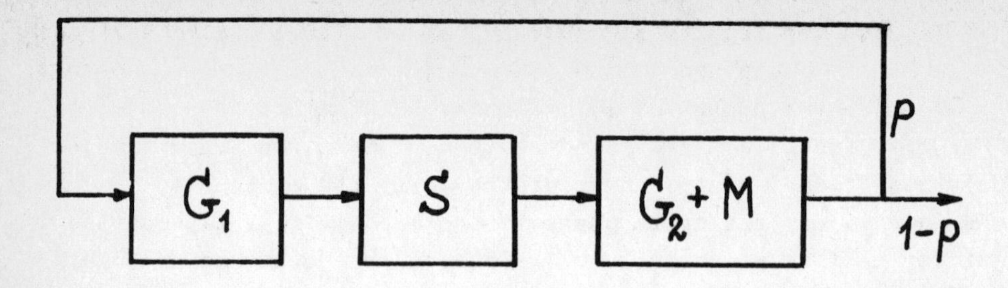

Figure 5.1. Block diagram of a simulation model for studying the statistical properties of cell kinetics indices.

and the solution to this equation is

$$q(t) = \mathbf{E}\left\{N(t)\right\} + \int_0^t \mathbf{E}\left\{(N(t-u)\right\} dR(u), \tag{13}$$

where $R(t) = \sum_{n=1}^{\infty} F^{*n}(t)$.

If $F(t)$ is Γ-distribution with a form parameter α and a scale parameter β for formula (13) we obtain

$$q(t)=\mathbf{E}\left\{N(t)\right\} + \sum_{n=1}^{\infty} \frac{\beta^{\alpha n}}{\Gamma(\alpha n)} \int_0^t \mathbf{E}\left\{N(t-u)\right\} e^{-\beta u} u^{\alpha n-1} du , \tag{14}$$

where $\alpha/\beta = \bar{\tau}$ is the average duration of the cycle phase and $\sqrt{\alpha}/\beta = \sigma$ is the standard deviation of the phase duration. In deriving (14) use was made of the fact of uniform convergence of the series

$$\sum_{n=1}^{\infty} \frac{t^{\alpha n-1}}{\Gamma(\alpha n)}$$

within every finite interval $[0, t_0]$.

In this case the variance of the integral cell flow into the cycle phase cannot be calculated directly on the basis of the process model, as one could do in describing a closed population by the methods of the theory of branching processes (see formula 10), it has to be evaluated by means of the sampling method. However, in an actual biological experiment that flow is an unobservable variable. It is apparent that, when applying in practice formula (14), the expectation $\mathbf{E}\left\{N(t)\right\}$ is substituted by its estimator, e.g. sampling mean value $\bar{N}(t)$ from realization of the $N(t)$ process. The variance of the estimate, in turn, is evaluated from direct observations of $N(t)$. Thus, expression (14) can be justly regarded as an estimator of the mathematical expectation of an integral cell flow into a transitive population (separate cycle phase) which is constructed from the experimentally observed N(t) function. Suppose q(t) is an estimator of the true q-index, in that case

$$q(t)=\bar{N}(t)+\sum_{n=1}^{\infty}\frac{\beta^{\alpha n}}{\Gamma(\alpha n)}\int_0^t \bar{N}(t-u)e^{-\beta u}u^{\alpha n-1}du \qquad (15)$$

Here we assume to know the values of the temporal parameters of the cycle phase with exhaustive precision.

It is obvious that, due to the independence and equal distribution of the values N(t) for each given instant t, the q(t) is a consistent and unbiased estimator for the q-index. Apparently, the simplest approach to studying the variance of this estimator is by means of computer simulation which admits direct recording both of the integral cell flow into a population and of the number of cells in the population at any specified instant t. Simulation provides a means for elucidating the following questions.

1. How closely does the q-index calculated by equation (14) from a single concrete realization of the N(t) process (let us denote it by $\tilde{q}(t)$) correspond to the observable index

$$q_{obs}(t) = \int_0^t K_1(\tau)d\tau \;?$$

2. How closely does the sampling mean $\bar{q}(t)$ of the integral cell flow $q_{obs}(t)$ is approximated by the estimator $\hat{q}(t)$ determined by formula (15)?

3. How does the variability of the N(t) process contribute to the variability of the variance of the q-index?

4. How inaccurate is information on the true q-index evaluated by formula (15) with inaccurately estimated temporal parameters τ and σ or incorrectly chosen parametric family of distributions which characterize the random duration of cell residence in a transitive population?

The last of the above questions is related to the problem of the robustness of the mathematical model which defines the procedure of constructing the q-index estimator. With a view to exploring the problems raised in the foregoing, experiments were conducted with a simulation model realizing the block diagram shown in Figure 5.1. The simplicity of the scheme does not restrict to any appreciable extent the generality of the investigation, since the main condition - the formation of a stochastic cell inflow - is fully satisfied within this scheme. It should be noted that some of the experiments listed below were reproduced with similar results on the more intricate models described in a paper by Zorin et al.[16] and in Chapter II of this book.

The simulation model of the system represented by the block diagram in Figure 5.1. was implemented with the aid of the GPSS/360 programming language. The S-phase was chosen as the cycle phase under study. In all the experiments described below, at the instant $t=0$, 10^3 cells were placed at the starting point of the G_1 phase, the p parameter was assumed to be equal to unity, and the values of the $q_{obs}(t)$ and $N(t)$ were recorded at hourly intervals.

In all the sets of experiments use was made of the simulation model with the following temporal parameters of cycle phases (hours):

$$\bar{\tau}_{G_1}=12, \sigma_{G_1}=2; \bar{\tau}_s=10, \sigma_s=2.5; \bar{\tau}_{G_2+M}=4, \sigma_{G_2+M}=0.54 .$$

Random durations of the G_1 and G_2+M phases were generated in the model in accordance with normal distribution and that of the S-phase with the Γ-distribution.

<u>Simulation experiment 1</u>. The experimental design was as follows: the q-index estimator was constructed from one realization of the $N(t)$ process, using formula (15) with exact $\bar{\tau}$ and σ values, and compared to the integral cell flow $q_{obs}(t)$ observable in the same experiment. Such comparisons were made for 100 realizations of the simulation model, and the maximum discrepance between the q-index estimator from a single realization $q(t)$ and the $q_{obs}(t)$ value in the time interval of up to 56 hours (with the average mitotic cycle duration of 26 hours) did not exceed 0.25 percent. Figure 5.2. presents one realization $q_{obs}(t)$.

<u>Simulation experiment 2</u>. In this set of experiments the following quantities were calculated from simulation data:

sampling mean of the $N(t)$ function for each fixed t (average over $N_i(t)$ realization ordinates) from the formula

$$\bar{N}(t) = \sum_{j=1}^{r} N_j(t)/r$$

and sampling standard deviation of the $N(t)$ function from the formula

$$S_{N(t)} = \sqrt{\sum_{j=1}^{r}[N_j(t)-\bar{N}(t)]^2/r-1} ,$$

where r is the number of realizations, as well as similar characteristics of the $q_{obs}(t)$ process, i.e. $\bar{q}_{obs}(t)$ and $S_{q_{obs}}(t)$. Besides,

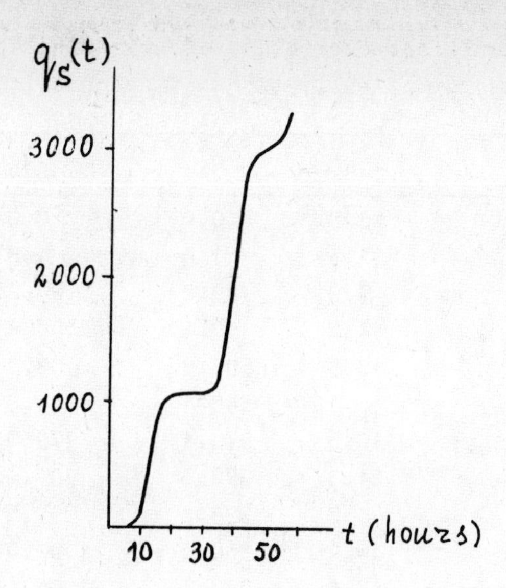

Figure 5.2. Realization of q-index obtained by computerized simulation of the layout in Figure 5.1.

using the $\bar{N}(t)$ values, $\hat{q}(t)$ estimator was computed from formula (15). The results of calculating these statistics for different numbers (5 and 20) simulation model realizations are given in Tables 5.1 and 5.2. These results indicate that given exact parametric information on the distribution of the duration of the cycle phase under study, the $\hat{q}(t)$ estimate provides a good approximation to the arithmetic mean of direct q-index observations. Indeed, the $\hat{q}(t)$ estimator of the actual index cannot be as accurate as that obtained through a direct averaging of the $q_{obs}(t)$ values.

This, in fact, is quite evident from the values of the estimators and of standard deviation for various stages of the evolution of a cell population presented in Table 5.1.

Simulation experiment 3. The purpose of this series of experiments was to ascertain how sensitive was the procedure for constructing the $\hat{q}(t)$ estimator to the choice of a concrete type of phase duration distribution and to errors in determining its numerical parameters, i.e. $\bar{\tau}$ and σ .

In the first variant of the model experiment the estimator $\hat{q}(t)$ for the S-phase of the mitotic cycle was calculated, as before, from formula (15) which is valid for the Γ-distribution of this phase duration. However, the duration of the S-phase in the simulation mo-

TABLE 5.1. Statistical characteristics of cell kinetic indices constructed from 5 realizations of stochastic processes.

t (hours)	N_s	Ns	q_s	$\bar{q}_s$	qs
4	0.0	0.0	0.0	0.0	0.0
6	1.8	1.5	1.8	1.8	1.5
8	23.2	7.0	23.2	23.2	7.0
10	159.4	11.8	159.4	159.4	11.8
12	502.4	18.3	502.8	503.0	18.0
14	841.2	6.1	845.1	844.8	7.8
16	953.0	10.6	978.1	979.0	5.1
18	889.8	21.3	992.8	997.4	1.1
20	717.8	24.8	993.3	999.8	0.4
22	469.4	16.1	968.2	1000.0	0.0
24	242.6	8.4	983.7	1000.4	0.5
26	109.6	5.6	996.3	1000.2	0.8
28	43.6	6.2	999.0	1006.2	4.0
30	51.2	11.9	1030.6	1038.2	9.8
32	128.8	28.0	1115.4	1125.4	27.0
34	327.4	28.5	1318.3	1327.4	27.4
36	649.4	34.7	1647.0	1655.6	34.0

del was generated according to normal distribution truncated on the left at point x=0. All the temporal parameters of the mitotic cycle phases (the S phase included) were the same as in experiments 1 and 2. The results of studying the robustness of the method for q-index evaluation to the choice of approximating distribution of phase duration are shown in Figure 5.3. These results indicate that evaluation of the $q_s(t)$ index from formula (15) is free from significant errors if the actual duration of the S phase is a normally distributed variable. This conclusion amplifies the existing concepts that different standard unimodal distributions of phase durations of the mitotic cycle, when used for analyzing cell kinetics curves, yield, as a rule, practically the same results [3, 6].

Another variant of the simulation experiment was concerned with the effect of errors made in determining numerical parameters $\bar{\tau}_s$ and σ_s of the S phase duration distribution on the procedure of constructing the q-index. In that case, both in the simulation model and in the algorithm of q-index estimation from the function $\bar{N}(t)$, there

TABLE 5.2. Statistical characteristics of cell kinetic indices constructed from 20 realizations of stochastic processes.

t	$\bar{N}_s$	σ_{N_s}	$\hat{q}_s$	$\bar{q}_s$	σ_{q_s}
4	0.0	0.0	0.0	0.0	0.0
6	1.2	1.0	1.2	1.2	1.0
8	21.6	5.1	21.6	21.6	5.1
10	161.2	13.4	161.2	161.2	13.4
12	501.4	12.6	501.7	501.7	12.5
14	842.5	10.3	845.9	845.6	10.5
16	953.2	6.7	977.7	977.7	4.5
18	893.5	12.5	996.3	998.3	1.3
20	722.3	15.9	998.0	999.9	0.3
22	478.4	15.2	995.6	1000.0	0.0
24	249.8	11.2	992.1	1000.3	0.4
26	106.1	7.9	995.5	1001.2	0.9
28	42.9	4.9	1002.9	1006.4	42.9
30	49.3	8.5	1035.1	1037.1	8.0
32	128.4	16.3	1121.8	1124.6	16.1
34	320.7	19.7	1317.0	1321.0	19.7
36	644.3	25.7	1646.6	1649.7	25.6

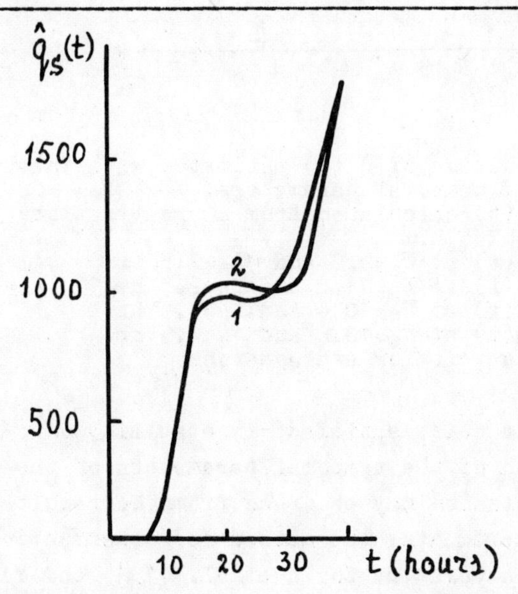

Figure 5.3. Simulation experiment 3. 1-$\bar{q}(t)$ at normal distribution of phase S duration (according to 20 realizations); 2 - $\hat{q}_s(t)$ estimator. See the text for explanations.

appeared the Γ-distribution of the duration of the S-phase. The results presented in Figure 5.4 show that the evaluation of the q-index from formula (15) is much more sensitive to inaccuracies in the assessment of the average value than of the standard deviation of the S-phase duration. We believe, at the same time, that these results

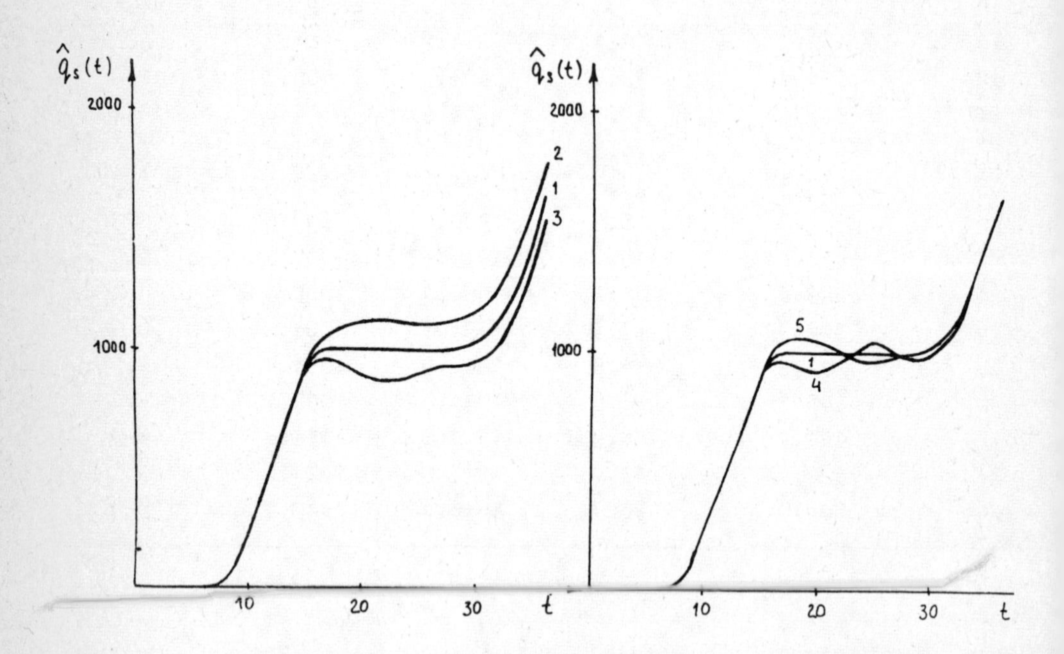

Figure 5.4. Construction of $\hat{q}_s(t)$ estimator with inaccurately defined phase S temporal parameters.
1 - $\bar{q}_s(t)$ calculated from 20 realizations at $\bar{\tau}_s$=10.0 and σ_s=2.5 hr;
2 - $\hat{q}_s(t)$ at $\bar{\tau}_s$=9.0 and σ_s=2.5 hr;
3 - $\hat{q}_s(t)$ at $\bar{\tau}_s$=11.0 and σ_s=2.5 hr;
4 - $\hat{q}_s(t)$ at $\bar{\tau}_s$=10.0 and σ_s=1.5 hr;
5 - $\hat{q}_s(t)$ at $\bar{\tau}_s$=10.0 and σ_s=3.5 hr.
See the text for explanations

bear witness to the quite satisfactory stability of q-index estimation to the variation of the temporal parameters of the cycle phase. The following conclusion may be drawn from the results of all the sets of simulations experiments: the method for constructing q-index estimator proposed in works by Zorin et al. [15] and Yakovlev at al. [14] features satisfactory accuracy and stability as regards incomplete parametric information on the distribution of the duration of the cycle phase under study. The situations is somewhat more involved in the case of estimating on the basis of radioautographical data the q-index normalized to the initial total number of cells in the population. In that case additional quantities - the mitotic index and the mean duration of mitotis - appear in the formula for calculating $q_s^0(t)$ estimator. As shown in our paper 14 , the behaviour of such q-index depends appreciably on the mean duration of mitosis, which imposes stringent requirements upon the accuracy of estimating that quantity from experimental data.

It should be noted, in conclusion, that certain features of the q-index bahaviour, specifically, failure of its property not to decrease monotonically for the period of observation, may serve as indications of errors in determining temporal parameters, inadequate experimental procedure or faulty interpretation of experimental evidence (see Figure 5.4, cf. [15]).

5.3. The Fraction of Labelled Mitoses Curve

The analysis of the fraction of labelled mitoses (FLM) curve is one of the most frequently used methods for estimating indirectly the numerical characteristics (mean and variance) of the lengths of the separate phases in the mitotic cycle. Nearly all experimental data on cell population kinetics whether "in vitro" or "in vivo" are interpreted by analysing the structure of FLMs. In the majority of such studies graphical methods are used for the estimation of the mitotic cycle phase lengths; these are not based on modern dynamic theory of cell systems. In order to obtain more sound methods it is necessary to construct a mathematical model for FLM curve and an appropriate procedure for non-linear estimation of the mitotic cycle temporal parameters. The basic principle for the indirect estimation of these parameters consists, therefore, of finding a set of their values which maximize,in some sense, the agreement between the model and the experimentally found FLM. The methods for describing the FLM curve mathematically which have been developed up to the present [see for review: 1, 3, 7] are usually applicable only to cell populations with stationary cell age (in respect to the starting point of the mitotic cycle) distribution - such populations either grow exponentially or are in the steady-state of growth. The works of MacDonald [6] and Jagers [4] contain the fullest theoretical solution for the FLM problem under such conditions. In experimental studies, however, much more complex situations can be tackled by analysing the FLM curve, e.g. diurnal variation in cell proliferation processes, transient cell kinetics in systems with induced DNA synthesis or in irradiated tissues. That is why the mathematical and computer simulation methods in the analysis of FLM curves deserve further development.

The simulation models discussed in Chapters II and IV may be used in investigating the peculiarities of the FLM pattern in different states of cell proliferation kinetics. In taking advantage of this possibility we proceeded from the following assumptions:

1. A labelled DNA precursor (^{3}H-thymidine) is included uniformly by all cells which are in the S-phase of the MC at the time of labelling regardless of the subpopulation the cells belong to.

2. The inclusion of a labelled precursor does not affect the cell's progress through its life cycle.

3. The number of labelled mitoses is recorded at the instant a mitosis terminates at each simulation step.

4. Radioactive label (tracer) is uniformly destributed among labelled-cell descendants.

Plotting the FLM curve on the basis of the foregoing assumptions can be readily done by means of the same software that was used in constructing the simulation model of cell population kinetics. Figure 5.5 presents a FLM curve obtained by averaging over 5 simulation runs conducted with

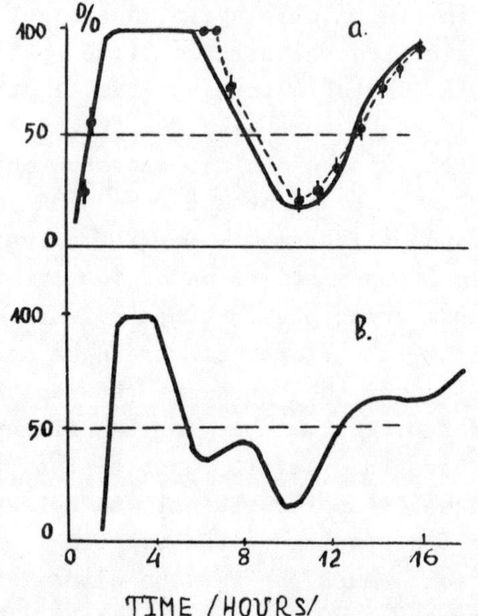

Figure 5.5. Changes in the FLM curve following ^{3}H thymidine administration 12 hours after a single irradiation. a - results of FLM simulation without irradiation; b - FLM simulated with irradiation in a dose of 6 Gy.

the model for the steady-state "crypt-villus" system discussed in Chapter IV. Presented in the same figure is an experimental FLM curve borrowed from [5]. In another simulation experiment irradiation-induced changes in the shape of FLM curves were investigated. For instance, Figure 5.5b shows a FLM curve (averaged over 5 realizations) obtained in simulating ^{3}H- thymidine administration 12 hours after irradiation with a dose of 6 Gy. The mean mitotic cycle duration calculated graphically from that curve, $\approx$10.6 h, is below the stationary value (13h). It should be emphasized that it was the stationary value of the mean mitotic cycle duration that was prescribed in simulating radiation inactivation of cells.

Thus, the only cause responsible for the distorted shape of the FLM curve obtained in simulation is the non-stationary rate of the cell transit through the mitotic cycle in the course of intestinal epithelium repopulation.

At the same time, similar alterations in the shape of FLM curves were also observed experimentally [5]. The results of simulation cast some doubt on the validity of the hypothesis advanced in [5] of the specific response of proliferating enterocytes to irradiation which manifests itself in a considerable decrease in the mean generation time. In effect, the observed changes in the FLM curve shape may occur with invariant temporal parameters of the cycle. They are due to the transient process taking place in the kinetics of irradiated cell populations of the intestinal epithelium.

The principal mechanism of intestinal epithelium repopulation after a single irradiation of mice with a dose of 6 Gy comprises the expansion of the proliferation zone in the crypts, induction of proliferative processes in the population of resting stem enterocytes and radiation damage repair of cells. These conclusions follow from experiments with the model described in Chapter IV.

REFERENCES

1. Eisen, M. Mathematical models in cell biology and cancer chemotherapy, Springer-Verlag, Berlin, Heidelberg, New York, 1979.
2. Harris, T.E. The theory of branching processes, Springer-Verlag, Berling-Göttingen-Heidelberg, 1963.
3. Hartmann, N.R., Gilbert, C.M., Jansson, B., Macdonald, P.D.M, Steel, G.G. and Valleron, A.J. A comparison of computer methods for the analysis of fraction labelled mitoses curves, Cell Tissue Kinet., 8, 119-124, 1975.
4. Jagers, P. Branching processes with biological applications, John Wiley and Sons, New York, 1975.

5. Lesher, S., Cooper, J., Hagemann, R. and Lesher, J. Proliferative patterns in the mouse jejunal epithelium after fractionated abdominal X-irradiation, Current Topics in Radiat. Res., 10, 229-261, 1975.
6. Machdonald, P.D.M. Statistical inference from the fraction labelled mitoses curve, Biometrika, 57, 489-503, 1970.
7. Malinin, A.M. and Yakovlev, A.Yu. The fraction of labelled mitoses curve in different states of cell proliferation kinetics. I. Common principles of a mathematical model, Cytology, 18, 1270-1277, 1976 (In Russian).
8. Mode, C.J. Multitype age-dependent branching processes and cell cycle analysis, Math. Biosci., 10, 177-190, 1971.
9. Sevastyanov, B.A. Branching processes, Nauka, Moscow, 1971 (In Russian).
10. Yakovlev, A.Yu. The dynamic reserving of hepatocytes as a mechanism of maintenance of regenerating liver specialized functions, Cytology, 21, 1243-1252, 1979 (In Russian).
11. Yakovlev, A.Yu., Malinin, A.M., Terskikh, V.V. and Makarova, G.F. Kinetics of induced cell proliferation at steady-state conditions of cell culture, Cytobiologie, 14, 279-283, 1977.
12. Yakovlev, A.Yu. and Yanev, N.M. The dynamics of induced cell proliferation within the model of branching stochastic process. I. Numbers of cells in successive generations, Cytology, 22, 945-953, 1980 (In Russian).
13. Yanev, N.M. and Yakovlev, A.Yu. The dynamics of induced cell proliferation systems within the model of a branching stochastic process. II. Some characteristics of the cell cycle temporal organization, Cytology, 25, 818-826, 1983 (In Russian).
14. Yakovlev, A.Yu., Zorin, A.V. and Isanin, N.A. The kinetic analysis of induced cell proliferation, J. Theor. Biol., 64, 1-25, 1977.
15. Zorin, A.V., Isanin, N.A. and Yakovlev, A.Yu. A study of the kinetics of transition of cells to DNA synthesis in regenerating liver, Acad. Sci. USSR Reports, 221, 226-228, 1973 (In Russian)
16. Zorin, A.V., Gushchin, V.A., Stefanenko, F.A., Gherepanova, O.N. and Yakovlev, A.Yu. Computer simulation of kinetics of irradiated cell populations in tumours, Exper. Oncology, 5, 23-30, 1983 (In Russian).

CONCLUSION

In recent years simulation techniques have been used to an increasing extent in describing a variety of biological processes, cell population kinetics in particular. This tendency is easily explicable. Indeed, creation of efficient means for interpreting experimental observations is inexorably associated with the development of models meeting stringent requirements for realistic description of the processes under study. Simulation modelling offers possibilities for conceptual analysis and prediction of reactions of complex systems with the minimum of limitations on the structure of the model and, in some cases, with replacing labour consuming biological experiments by their simulation on a computer.

In the present book primary emphasis is placed on simulation of the effects of cellular radiobiology. This is explained not only by the professional interests of the authors but also by the fact that radiobiology is one of the branches of present-day biological science most fit to make use of mathematical methods. Practically from its outset at the turn of the century and up to the present radiobiology has been developing on a quantitative basis. A particularly important part in quantitative radiobiology is played by probabilistic models, seeing that the stochastic nature of both radiation effect and the response of the biological system has been long recognized by radiobiologists as a fact requiring no special proof.

The literature on different models of dose-effect ralationship for homogeneous cell populations is quite voluminous. Such models are essentially descriptive, their purpose being to establish the relationship between probability of cell survival (estimated by clonogenic capacity of cells as observed in experiment) and the physical characteristics of irradiation. They appear to be of little use for gaining more intimate knowledge of the factors determining the peculiarities of specific radiobiological effects in nonhomogeneous cell populations of dissimilar radiosensitivity. Attempts to describe in detail within a framework of a probabilistic model the effect of numerous factors modifying cell response to irradiation (such attempt have been made involving, as a rule, the use of models of the type of Markovian stochastic processes) run into apparent mathematical difficulties. Transition to direct computer simulation of cellular radiobiology phenomena, based on regarding an irradiated cell population as a discrete stochastic system, offers distinct advantages. It should be remebered, however, that such a transition is attended by the loss of possibilities intrinsic in analytical study of a model. It might be well to seek a reasonable compromise between the two approaches, supplementing and enriching on a reciprocal basis conclusions that ensue from their application.

The present volume deals in detail with two stochastic simulation models oriented to studying the effect of ionizing radiation on mammalian cell culture and the intestinal epithelium. Outlined below are some of the principal results obtained from the application of the models to the analysis of concrete experimental observations:

- practically all known effects of cellular radiobiology have been reproduced with the cell culture simulation model, which lends support to the fundamental validity of present-day conceptions of regularities inherent in radiation injury and post-irradiation recovery of mammalian cells;
- for the first time quantitative estimates have been obtained of the radiosensitivity characteristics and of the ability of resting and proliferating in vitro cells to recover from radiation damage. It has also been found that cells in the resting state may exhibit at the same time both higher radiosensitivity and greater intensity of repair processes than their proliferating counterparts;
- explanation has been offered for dissimilarities in the response of radiocurable and radioincurable human tumour cells to fractionated irradiation;
- quantitative studies have been conducted of the contribution of certain biological characteristics of the cellular system to its response to irradiation;
- it has been demonstrated that a relative rise in survival with increase in irradiation dose may be accounted for in the stationary cell culture by the non-linear contribution to survival of the processes of radiation damage formation and repair rather than by higher intensity of repair processes;
- experimental data on the effect of radiation upon the small-intestine epithelium, one of the most radiosensitive systems of the organism, have been reproduced; the data have been interpreted from the standpoint of cell kinetics, and radiobiological characteristics of the mice small intestine have been identified;
- it has been found that changes in the form of the labelled mitoses curve in the course of post-irradiation tissue recovery may occur with invariant temporal parameters of the mitotic cycle of cells, the thesis of a substantial reduction in the average duration of the cycle

of irradiated enterocytes, commonly encountered in the literature, requiring a revision;

- new evidence has been obtained of the part played by the tissue regulation system in the post-irradiation recovery of the intestinal epithelium. It has been shown that the mechanisms maintaining tissue homeostasis are capable of ensuring not only the dynamic but also the stochastic stability of renewing cellular systems, which manifests itself in the limitation of variance of random fluctuations in the cell subpopulations numbers. Another interesting feature of regulated cellular systems is stabilization of the size of the population under the influence of external random perturbations affecting some elements of the tissue regulation system.

Another important field of application for simulation modelling in cell biology consists in investigating the characteristics of cell kinetics indicators and elucidating the limits of applicability of mathematical methods previously proposed for their analysis. In one of the chapters of the book we have attempted to demonstrate the potentialities of that particular area in simulating cellular systems.

There is no doubt that despite the very complex structure and the assimilation of a considerable body of up-to-date data the models discussed in the book are far from completion. It seems quite likely that certain modifications of the models would be required when employing them in studying the effects of complex dynamic regimens of fractionated irradiation or in exploring the relative biological efficiency of irradiation differing in quality. Such modifications would also appear inevitable when passing to simulation of other cellular systems. However, the investigator's endeavours in revising or refining elements of a model which represents some biological phenomenon will not be in vain. They are rewarded with a new understanding of the essence of the processes under study which, indeed, is the primary aim of scientific research.

Bio-mathematics

Managing Editor: **S. A. Levin**

Volume 15
D. L. DeAngelis, W. M. Post, C. C. Travis

Positive Feedback in Natural Systems

1986. 90 figures. XII, 290 pages. ISBN 3-540-15942-8

Contents: Introduction. - The Mathematics of Positive Feedback. - Physical Systems. - Evolutionary Processes. - Organisms Physiology and Behavior. - Resource Utilization by Organisms. - Social Behavior. - Mutualistic and Competitive Systems. - Age-Structured Populations. - Spatially Heterogeneous Systems: Islands and Patchy Regions. - Spatially Heterogeneous Ecosystems; Pattern Formation. - Disease and Pest Outbreaks. - The Ecosystem and Succession. - Appendices. - References. - Subject Index. - Author Index.

Volume 16

Complexity, Language, and Life: Mathematical Approaches

Editors: **J. L. Casti, A. Karlqvist**

1986. XIII, 281 pages. ISBN 3-540-16180-5

Contents: Allowing, forbidding, but not requiring: a mathematic for human world. - A theory of stars in complex systems. - Pictures as complex systems. - A survey of replicator equations. - Darwinian evolution in ecosystems: a survey of some ideas and difficulties together with some possible solutions. - On system complexity: identification, measurement, and management. - On information and complexity. - Organs and tools; a common theory of morphogenesis. - The language of life. - Universal principles of measurement and language functions in evolving systems.

Volume 17

Mathematical Ecology

An Introduction

Editors: **Th. G. Hallam, S. A. Levin**

1986. 84 figures. XII, 457 pages. ISBN 3-540-13631-2

Contents: Introduction. - Physiological and Behavioral Ecology. - Population Ecology. - Communities and Ecosystems. - Applied Mathematical Ecology. - Author Index. - Subject Index.

Volume 18

Applied Mathematical Ecology

Editors: **S. A. Levin, T. G. Hallam, L. J. Gross**

1988. ISBN 3-540-19465-7. In preparation.

Volume 19
J. D. Murray

Mathematical Biology

1988. ISBN 3-540-19460-6. In preparation.

Springer-Verlag
Berlin Heidelberg New York
London Paris Tokyo

Lecture Notes in Biomathematics

Vol. 58: C. A. Macken, A. S. Perelson, Branching Processes Applied to Cell Surface Aggregation Phenomena. VIII, 122 pages. 1985.

Vol. 59: I. Nåsell, Hybrid Models of Tropical Infections. VI, 206 pages. 1985.

Vol. 60: Population Genetics in Forestry. Proceedings, 1984. Edited by H.-R. Gregorius. VI, 287 pages 1985.

Vol. 61: Resource Management. Proceedings, 1984. Edited by M. Mangel. V, 138 pages 1985.

Vol. 62: B. J. West, An Essay on the Importance of Being Nonlinear. VIII, 204 pages. 1985.

Vol. 63: Carme Torras i Genís, Temporal-Pattern Learning in Neural Models. VII, 227 pages, 1985.

Vol. 64: Peripheral Auditory Mechanisms. Proceedings, 1985. Edited by J.B. Allen, J.L. Hall, A. Hubbard, S. Neely and A. Tubis. VII, 400 pages. 1986.

Vol. 65: Immunology and Epidemiology. Proceedings, 1985. Edited by G.W. Hoffmann and T. Hraba. VIII, 242 pages. 1986.

Vol. 66: Nonlinear Oscillations in Biology and Chemistry. Proceedings, 1985. Edited by H.G. Othmer. VI, 289 pages. 1986.

Vol. 67: P.H. Todd, Intrinsic Geometry of Biological Surface Growth. IV, 128 pages. 1986.

Vol. 68: The Dynamics of Physiologically Structured Populations. Edited by J.A.J. Metz and O. Diekmann. XII, 511 pages. 1986.

Vol. 69: S.H. Strogatz, The Mathematical Structure of the Human Sleep-Wake Cycle. VIII, 239 pages. 1986.

Vol. 70: Stochastic Methods in Biology. Proceedings, 1985. Edited by M. Kimura, G. Kallianpur and T. Hida. VI, 229 pages. 1987.

Vol. 71: Mathematical Topics in Population Biology, Morphogenesis and Neurosciences. Proceedings, 1985. Edited by E. Teramoto and M. Yamaguti. IX, 348 pages. 1987.

Vol. 72: Modeling and Management of Resources under Uncertainty. Proceedings, 1985. Edited by T.L. Vincent, Y. Cohen, W.J. Grantham, G.P. Kirkwood, and J.M. Skowronski. VIII, 318 pages. 1987.

Vol. 73: Y. Cohen (Ed.), Applications of Control Theory in Ecology. Proceedings, 1986. VII, 101 pages. 1987.

Vol. 74: A.Y. Yakovlev, A.V. Zorin, Computer Simulation in Cell Radiobiology. VI, 133 pages. 1988.

This series reports new developments in biomathematics research and teaching – quickly, informally and at a high level. The type of material considered for publication includes:

1. Original papers and monographs
2. Lectures on a new field or presentations of new angles in a classical field
3. Seminar work-outs
4. Reports of meetings, provided they are
   a) of exceptional interest and
   b) devoted to a single topic.

Texts which are out of print but still in demand may also be considered if they fall within these categories.

The timeliness of a manuscript is more important than its form, which may be unfinished or tentative. Thus, in some instances, proofs may be merely outlined and results presented which have been or will later be published elsewhere. If possible, a subject index should be included. Publication of Lecture Notes is intended as a service to the international scientific community, in that a commercial publisher, Springer-Verlag, can offer a wide distribution of documents which would otherwise have a restricted readership. Once published and copyrighted, they can be documented in the scientific literature.

Manuscripts

Manuscripts should be no less than 100 and preferably no more than 500 pages in length.
They are reproduced by a photographic process and therefore must be typed with extreme care. Symbols not on the typewriter should be inserted by hand in indelible black ink. Corrections to the typescript should be made by pasting in the new text or painting out errors with white correction fluid. The typescript is reduced slightly in size during reproduction; best results will not be obtained unless on each page a typing area of 18 × 26.5 cm (7 × 10½ inches) is respected. On request the publisher can supply paper with the typing area outlined. Move detailed typing instructions are also available on request.

Manuscripts generated by a word-processor or computerized typesetting are in principle acceptable. However if the quality of this output differs significantly from that of a standard typewriter, then authors should contact Springer-Verlag at an early stage.

Authors of monographs and editors of proceedings receive 50 free copies.

Manuscripts should be sent to Prof. Simon Levin, Section of Ecology and Systematics, 345 Corson Hall, Cornell University, Ithaca, NY 14853-0239, USA, or directly to Springer-Verlag Heidelberg.

Springer-Verlag, Heidelberger Platz 3, D-1000 Berlin 33
Springer-Verlag, Tiergartenstraße 17, D-6900 Heidelberg 1
Springer-Verlag, 175 Fifth Avenue, New York, NY 10010/USA
Springer-Verlag, 37-3, Hongo 3-chome, Bunkyo-ku, Tokyo 113, Japan

---

ISBN 3-540-19457-6
ISBN 0-387-19457-6

RANDALL LIBRARY-UNCW
3 0490 0087970 4